Implementing IP and Ethernet on the 4G Mobile Network

Implementing IP and Ethernet on the 4G Mobile Network

André Perez

ISTE PRESS

ELSEVIER

First published 2017 in Great Britain and the United States by ISTE Press Ltd and Elsevier Ltd

ISTE Press Ltd
27-37 St George's Road
London SW19 4EU
UK

www.iste.co.uk

Elsevier Ltd
The Boulevard, Langford Lane
Kidlington, Oxford, OX5 1GB
UK

www.elsevier.com

British Library Cataloguing-in-Publication Data
A CIP record for this book is available from the British Library
Library of Congress Cataloging in Publication Data
A catalog record for this book is available from the Library of Congress
ISBN 978-1-78548-238-0

Contents

Preface

This book deals with the combination of 4G mobile network and transport network of Internet Protocol (IP) packets and Ethernet frames through the analysis of data transfer functions (Chapters 1–6), quality of service (Chapters 7–9), security (Chapters 10–12) and synchronization (Chapters 13–15).

Mobile network

The 4G mobile network allows the flow (IP packet) of the mobile to be transported to its PDN Gateway (PGW) and vice versa.

The flow (IP packet) of the mobile is transported by bearers that are built between the various entities of the 4G mobile network (Figure P.1):

– Data Radio Bearer (DRB) built between the User Equipment (UE) and the evolved Node Base (eNB) station;

– S1 bearer built between eNB and Serving Gateway (SGW) entities;

– S5 bearer built between SGW and PGW entities.

The IP packet, related to the S1 or S5 bearer, contains the IP packet, related to the mobile flow.

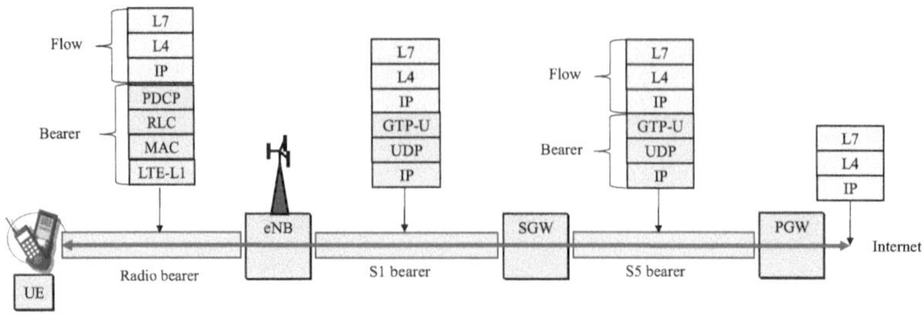

Figure P.1. *Mobile flow transport in a bearer*

Transport network

The transport network of IP packets or Ethernet frames can build a network interconnection for the various entities of the 4G mobile network and consists of the following networks (Figure P.2):

– Multiprotocol Label Switching-Virtual Private Network (MPLS-VPN) allows us to build an IP network to interconnect different entities of the 4G mobile core network: Mobility Management Entity (MME), SGW, PGW, Home Subscriber Server (HSS) and Policy and Charging Rule Function (PCRF);

– Virtual Private LAN Service (VPLS) allows us to build an Ethernet network for interconnection of eNB entities to the 4G mobile core network (MME, SGW);

– Optical Transport Network (OTN) is a transmission network over optical fiber of Ethernet frames for interconnection over long distances of the various entities Provider Edge (PE) and Provider (P) of the MPLS-VPN and VPLS networks.

S1 bearer transport

The transport of IP packets, related to S1 bearer, is provided by Ethernet frames in the VPLS network between the eNB entity and the R1 router, then by Ethernet frames in the LAN1 network between the R1 router and the SGW entity (Figure P.3).

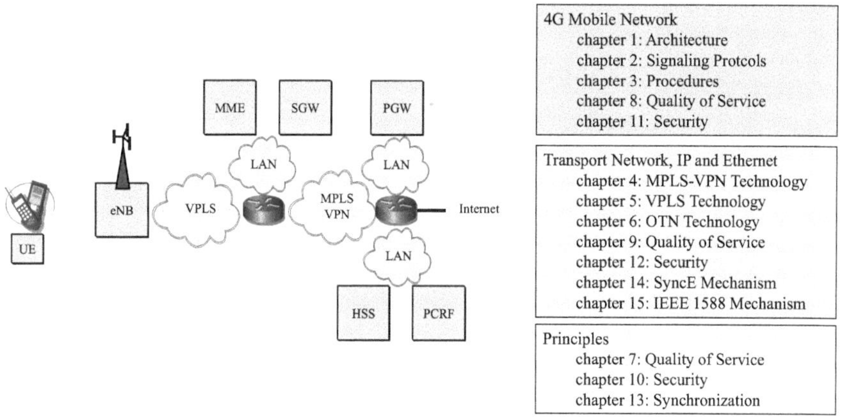

Figure P.2. *4G mobile network and transport network*

Ethernet frames into the VPLS network are switched by the PE entities and transported between the PE entities in virtual circuits, the P entity ensuring the label switching.

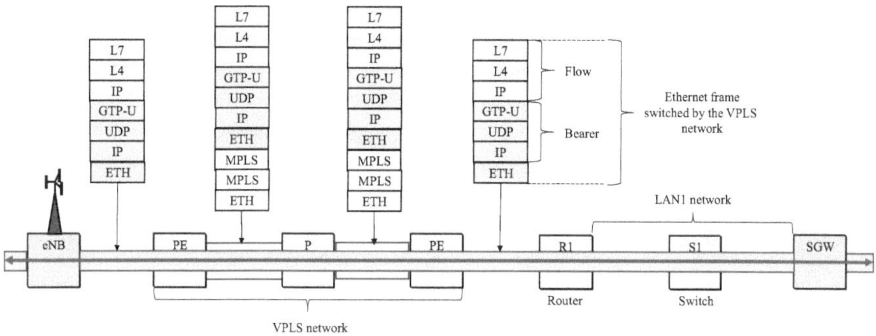

Figure P.3. *S1 bearer transport*

S5 bearer transport

The IP packets, related to the S5 bearer, are transported by Ethernet frames in the LAN1 network, between the SGW entity and the R1 router, and by Ethernet frames in the LAN2 network, between the R2 router and the PGW entity (Figure P.4).

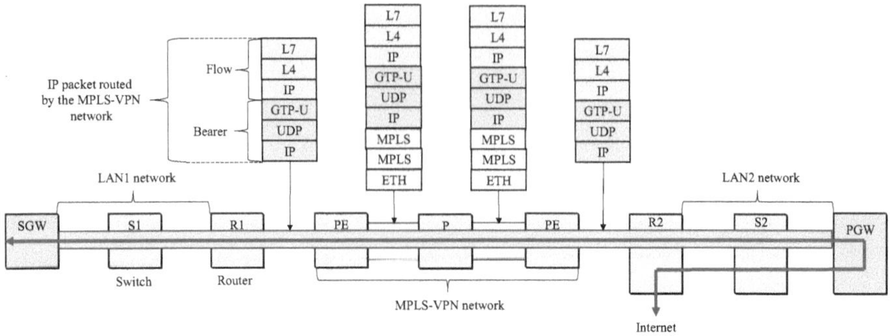

Figure P.4. *S5 bearer transport*

The IP packets, related to the S5 bearer, are routed by the R1 router, PE entities of the MPLS-VPN network and R2 router.

The IP packets into the MPLS-VPN network are transported between PE entities in virtual circuits, the entity P ensuring the label switching.

The IP packet, related to the mobile flow, is routed by the PGW entity, transported by Ethernet frames to the R2 router, and then routed by the R2 router to access the Internet network.

OTN network

OTN provides the constitution of the following components:

– Optical Channel (OCh) based on data from PE or P entities of the MPLS-VPN or VPLS networks;

– Optical Multiplex Section (OMS) performing the wavelength-division multiplexing of different OChs;

– Optical Transmission Section (OTS) of the wavelength multiplex (Figure P.5).

The structure of the OTN depends on the network topology and includes Optical Line Terminal for constitution of linear links, Optical Add-Drop Multiplexer (OADM) for constitution of linear or ring network or Optical Cross-Connect (OXC) for constitution of mesh networks.

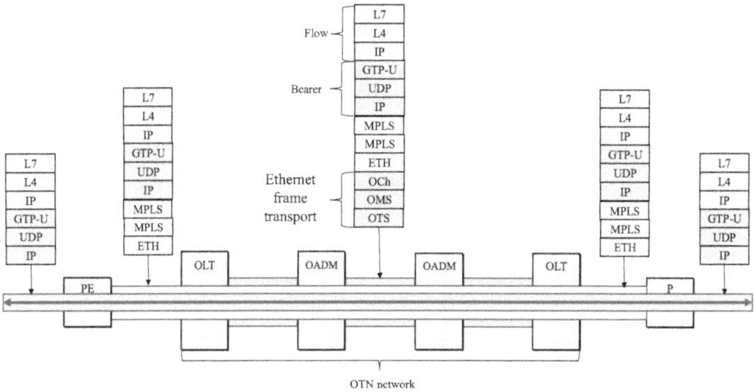

Figure P.5. *Optical transport network*

Quality of service

Flows and bearers are grouped into classes of service identified by the following parameters:

– QoS Class Identifier (QCI) for the bearers DRB, S1 and S5;

– DiffServ Code Point (DSCP) for the IP packets;

– Priority Code Point (PCP) for Ethernet frames;

– EXP or Traffic Class (TC) for labeled packets or frames.

The quality of service comprises on applying to each data structure the behavior (congestion avoidance, scheduling) based on the value of the identifier.

The value of the QCI parameter applied to a data structure is defined by the 4G mobile network. The value of other parameters is obtained by mapping from the QCI parameter.

Security

Security in 4G mobile network

The security architecture implemented in the 4G mobile network is on the attainment of the following:

– mutual authentication of the 4G network and the mobile;

– security of the signaling messages Non-Access Stratum (NAS) exchanged between the mobile and the MME entity. The security regards the integrity control and the encryption of the messages;

– security of the radio interface Long-Term Evolution-Uu (LTE-Uu) between the mobile and the eNB entity. The security regards, first, the integrity control and the encryption of the messages Radio Resource Control (RRC) and, second, the encryption of IP packets of the user plane.

Bearer protection

IP Security mechanism implements protection of S1 and S5 bearers between the following entities (Figure P.6):

– eNB entity and SEG1 (Security Gateway) entity located in the LAN1 network for S1 bearer;

– SEG1 and SEG2 entities located, respectively, in the LAN1 and LAN2 networks for S5 bearer.

The IP packet, related to the mobile flow, is protected at the Packet Data Convergence Protocol layer on the radio interface LTE-Uu (Figure P.6).

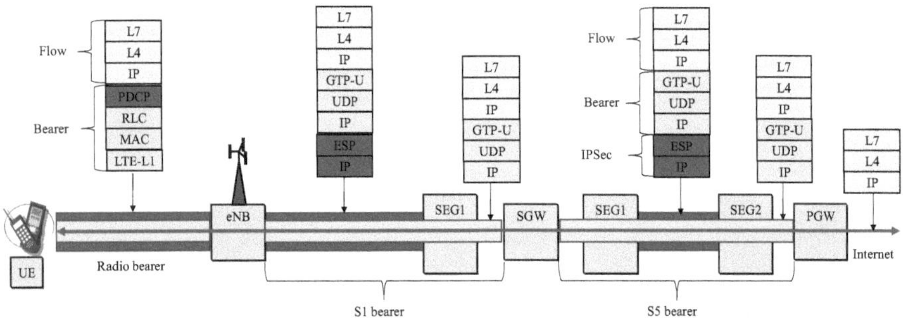

Figure P.6. *Bearer protection. For a color version of this figure, see www.iste.co.uk/perez/4gmobile.zip*

Synchronization

The different entities of the mobile network, with the exception of the eNB entity and the transport network, require no synchronization to implement the data transfer.

The radio interface of eNB entities must be synchronized in frequency to ensure the change-of cells during the session (handover) for both modes of operation, the Frequency Division Duplex (FDD) and Time Division Duplex (TDD).

For the TDD mode, time synchronization is required to synchronize the time of transmission and reception of eNB entities.

The interference management between eNB entities, based on time sharing of the radio resource, also requires time synchronization of the eNB entities.

The method based on a synchronous physical layer, implemented by the Synchronous Ethernet mechanism (SyncE), is a method that suits frequency synchronization.

The method based on packets used in bidirectional mode, implemented by the IEEE 1588 mechanism, allows time and frequency synchronization.

List of Abbreviations

A

AAA *Authenticate and Authorize Answer*

AAR *Authenticate and Authorize Request*

AC *Attachment Circuit*

AES *Advanced Encryption Standard*

AF *Application Function*

AF *Assured Forwarding*

AH *Authentication Header*

AIA *Authentication-Information-Answer*

AIR *Authentication-Information-Request*

AKA *Authentication and Key Agreement*

AM *Acknowledged Mode*

AMBR *Aggregate Maximum Bit Rate*

AMR *Adaptive Multi-Rate*

AMR WB *AMR Wide Band*

AP *Application Part*

APN *Access Point Name*

APS *Automatic Protection Switching*

ARP	*Allocation and Retention Priority*
ARQ	*Automatic Repeat request*
AS	*Autonomous System*
ASA	*Abort-Session-Answer*
ASR	*Abort-Session-Request*
AUTN	*Authentication Network*

B

BA	*Behavior Aggregate*
BC	*Boundary Clock*
BDI	*Backward Defect Indication*
BDI-O	*Backward Defect Indication – Overhead*
BDI-P	*Backward Defect Indication – Payload*
BEI	*Backward Error Indication*
BIAE	*Backward Incoming Alignment Error*
BIP-8	*Bit Interleaved Parity*
BMCA	*Best Master Clock Algorithm*

C

CA	*Certificate Authority*
CAC	*Connection Admission Control*
CBP	*Constrained Baseline Profile*
CBS	*Committed Burst Size*
CS/CB	*Coordinated Scheduling / Beamforming*
CCA	*Credit-Control-Answer*
CCR	*Credit-Control-Request*
CE	*Customer Edge*

CFI	*Canonical Format Indicator*
cHEC	*core Header Error Check*
CIR	*Committed Information Rate*
CLA	*Cancel-Location-Answer*
CLR	*Cancel-Location-Request*
CMAS	*Commercial Mobile Alert System*
CMP	*Certificate Management Protocol*
CoMP	*Coordinated Multi-Point operation*
CP	*Cyclic Prefix*
CRL	*Certificate Revocation List*
C-RNTI	*Cell Radio Network Temporary Identity*
CS	*Circuit-Switched*
CS	*Class Selector*
CSR	*Certificate Signing Request*
CWDM	*Coarse WDM*

D

DCM	*Dispersion Compensation Module*
DDA	*Delete-Subscriber-Data-Answer*
DDR	*Delete-Subscriber-Data-Request*
DEI	*Drop Eligible Indicator*
DF	*Don't Fragment*
DiffServ	*Differentiated Services*
DL	*downlink*
DLCI	*Data Link Connection Identifier*
DNS	*Domain Name Server*

DNU *Do Not Use*

DOI *Domain of Interpretation*

DPS *Dynamic Point Selection*

DRB *Data Radio Bearer*

DRR *Deficit Round Robin*

DSCP *DiffServ Code Point*

DWDM *Dense WDM*

DwPTS *Downlink Pilot Time Slot*

E

E2E *End to End* TC

ECGI *E-UTRAN Cell Global Identifier*

ECN *Explicit Congestion Notification*

EEA *EPS Encryption Algorithm*

EEC *Ethernet Equipment Clock*

EF *Expedited Forwarding*

EIA *EPS Integrity Algorithm*

E-LSP *EXP-inferred-class LSP*

EMM *EPS Mobility Management*

eNB *evolved Node Base station*

EoMPLS *Ethernet over MPLS*

EPC *Evolved Packet Core*

EPS *Evolved Packet System*

E-RAB *EPS Radio Access Bearer*

ESM *EPS Session Management*

ESMC *Ethernet Synchronization Messaging Channel*

ESP	*Encapsulating Security Payload*
ETWS	*Earthquake and Tsunami Warning System*
E-UTRAN	*Evolved Universal Terrestrial Radio Access Network*
EVS	*Enhanced Voice Services*
EXI	*Extension Header Identifier*

F

FAS	*Frame Alignment Signal*
FB	*Full Band*
FCS	*Frame Check Sequence*
FDI-O	*Forward Defect Indication – Overhead*
FDI-P	*Forward Defect Indication – Payload*
FDD	*Frequency Division Duplex*
FEC	*Forwarding Equivalence Class*
FEC	*Forward Error Correction*
FIB	*Forwarding Information Base*
FIFO	*First-In First-Out*
FTTL	*Fault Type and Fault Location*

G

GBR	*Guaranteed Bit Rate*
GCC	*General Communication Channel*
GFP	*Generic Framing Procedure*
GFP-F	GFP *Frame-mapped*
GFP-T	GFP *Transparent-mapped*

GP	*Gap Period*
GTP-U	*GPRS Tunnel Protocol User*
GTPv2-C	*GPRS Tunnel Protocol version 2 Control*
GUMMEI	*Globally Unique MME Identity*
GUTI	*Globally Unique Temporary Identity*

H

HARQ	*Hybrid Automatic Repeat request*
HeNB	*Home eNB*
HII	*High Interference Indication*
HMAC	*Keyed-Hash Message Authentication Code*
H-PCRF	*Home-PCRF*
HR	*Home Routed*
HSS	*Home Subscriber Server*
H-VPLS	*Hierarchical VPLS*

I, J, K

IAE	*Incoming Alignment Error*
ICV	*Integrity Check Value*
IDA	*Insert-Subscriber-Data-Answer*
IDR	*Insert-Subscriber-Data-Request*
IFG	*Inter-Frame Gap*
IKE	*Internet Key Exchange*
IMS	*IP Multimedia Sub-system*
IMSI	*International Mobile Subscriber Identity*
IOI	*Interference Overload Indication*
IP	*Internet Protocol*

IPSec	*IP Security*
ISAKMP	*Internet Security Association and Key Management Protocol*
IS-IS	*Intermediate System to Intermediate System*
ITU	*International Telecommunication Union*
IV	*Initialization Vector*
JR	*Joint Reception*
JT	*Joint Transmission*
KDF	*Key Derivation Function*

L

LAN	*Local Area Network*
LB	*Label Base*
LBO	*Local BreakOut*
LCID	*Logical Channel Identifier*
LDP	*Label Distribution Protocol*
LFIB	*Label Forwarding Information Base*
LIB	*Label Information Base*
LLQ	*Low Latency Queueing*
L-LSP	*Label-inferred-class LSP*
LSP	*Label Switching Path*
LSR	*Label Switching Router*

M

MAC	*Message Authentication Code*
MAC	*Medium Access Control*
MBR	*Maximum Bit Rate*
MBSFN	*MBMS over Single Frequency Network*

MC	*Matrix Connection*
MCC	*Mobile Country Code*
MD5	*Message Digest*
MFAS	*Multi-Frame Alignment Signal*
MME	*Mobility Management Entity*
MNC	*Mobile Network Code*
MODP	*MODular exponential modulus* P
MOS	*Mean Opinion Score*
MP	*Main Profile*
MP-BGP	*Multi-Protocol Border Gateway Protocol*
MPLS	*Multi-Protocol Label Switching*
MTBF	*Mean Time Between Failure*
MTIE	*Maximum Time Interval Error*
MTTR	*Mean Time To Repair*

N

NAPT	*Network Address and Port Translation*
NAS	*Non-Access Stratum*
NB	*Narrow Band*
NLRI	*Network Layer Reachability Information*
NOA	*Notify-Answer*
NOR	*Notify-Request*
N-PE	*Network-facing PE*

O

OA	*Optical Amplifier*
OADM	*Optical Add Drop Multiplexer*

OC	*Ordinary Clock*
OCh	*Optical Channel*
OCI	*Open Connection Indication*
ODU	*Optical channel Data Unit*
OFDM	*Orthogonal Frequency Division Multiplexing*
OH	*Over-Head*
OLT	*Optical Line Terminal*
OMS	*Optical Multiplex Section*
OPU	*Optical channel Payload Unit*
OSC	*Optical Supervisory Channel*
OSPF	*Open Shortest Path First*
OSSP	*Organization Specific Slow Protocol*
OTN	*Optical Transport Network*
OTS	*Optical Transmission Section*
OTU	*Optical channel Transport Unit*
OUI	*Organizationally Unique Identifier*
OXC	*Optical Cross-Connect*

P, Q

P	*Provider*
P2P	*(Peer to Peer) TC*
PAD	*Peer Authorization Database*
PBS	*Peak Burst Size*
PCC	*Policy and Charging Control*
PCC	*Protection Communication Channel*
PCEF	*Policy and Charging Enforcement Function*
PCM	*Pulse Code Modulation*
PCP	*Priority Code Point*

PCRF *Policy and Charging Rule Function*

PDCP *Packet Data Convergence Protocol*

PDN *Packet Data Network*

PDU *Protocol Data Unit*

PDV *Packet Delay Variation*

PE *Provider Edge*

pFCS *payload Frame Check Sequence*

PFI *Payload FCS Indicator*

PGW *PDN Gateway*

PHB *Per-Hop Behavior*

PHP *Penultimate Hop Popping*

PIR *Peak Information Rate*

PKI *Public Key Infrastructure*

PLC *Packet Loss Concealment*

PLI *Payload Length Indicator*

PLL *Phase-Locked Loop*

PM *Path Monitoring*

PMD *Polarization Modal Dispersion*

PMI *Payload Missing Indication*

PMTU *Path MTU*

PQ *Priority Queueing*

PRC *Primary Reference Clock*

PS *Packet-Switched*

PSI *Payload Structure Identifier*

PT *Payload Type*

PTI *Payload Type Identifier*

PTP *Precision Time Protocol*

PUA *Purge-UE-Answer*

PUR	*Purge-UE-Request*
PW	*Pseudo-Wire*
QCI	*QoS Class Identifier*
QL	*Quality Level*
QoE	*Quality of Experience*
QoS	*Quality of Service*

R

RA	*Registration Authority*
RAA	*Re-Auth-Answer*
RAR	*Re-Auth-Request*
RAT	*Radio Access Technology*
RD	*Route Distinguisher*
RED	*Random Early Discard*
RES	*Result*
RIB	*Routing Information Base*
ROHC	*Robust Header Compression*
RLC	*Radio Link Control*
RNTP	*Relative Narrowband Tx Power*
RR	*Reflector Route*
RR	*Round Robin*
RRC	*Radio Resource Control*
RSA	*Reset-Answer*
RSR	*Reset-Request*
RT	*Route Target*
RTP	*Real-time Transport Protocol*
RTT	*Round Trip Time*

S

SA	*Security Association*
SAD	*Security Association Database*
SDF	*Service Data Flow*
SDH	*Synchronous Digital Hierarchy*
SEG	*Security Gateway*
SFD	*Start of Frame Delimiter*
SFN	*System Frame Number*
SGW	*Serving Gateway*
SHA	*Secure Hash Algorithm*
SIP	*Session Initiation Protocol*
SLR	*Send Loudness Rating*
SM	*Section Monitoring*
SN	*Sequence Number*
SN ID	*Serving Network Identifier*
SONET	*Synchronous Optical Network*
SPD	*Security Policy Database*
SPI	*Security Parameters Index*
SPR	*Subscription Profile Repository*
sr TCM	*single rate Three Color Marker*
SRB	*Signaling Radio Bearer*
SSM	*Synchronization Status Message*
SSU	*Synchronization Supply Unit*
STA	*Session Termination Answer*
STP	*Spanning Tree Protocol*

STR	*Session Termination Request*
S-VLAN	*Service VLAN*
SWB	*Super Wide Band*
SyncE	*Synchronous Ethernet*

T

TAC	*Tracking Area Code*
TAI	*Tracking Area Identity*
TC	*Traffic Class*
TC	*Transparent Clock*
TCI	*Tag Control Identification*
TCM	*Tandem Connection Monitoring*
TCP	*Transmission Control Protocol*
TDD	*Time Division Duplex*
TDEV	*Time Deviation*
TEID	*Tunnel Endpoint Identifier*
TFT	*Traffic Flow Template*
tHEC	*type Header Error Check*
TIE	*Time Interval Error*
TL	*Tunnel Label*
TLV	*Type Length Value*
TM	*Transparent Mode*
ToS	*Type of Service*
TPID	*Tag Protocol Identification*
trTCM	*two rate Three Color Marker*
TS	*Time Stamp*
TTI	*Transmitted Trace Identifier*
TTL	*Time To Live*

U

UE	*User Equipment*
UICC	*Universal Integrated Circuit Card*
UIpp	*Unit Interval peak-to-peak*
UL	*uplink*
ULA	*Update-Location-Answer*
ULR	*Update-Location-Request*
UM	*Unacknowledged Mode*
U-PE	*User-facing PE*
UPI	*User Payload Identifier*
UpPTS	*Uplink Pilot Time Slot*
USIM	*Universal Services Identity Module*
UTC	*Universal Time Coordinated*

V

VBO	*VPLS endpoint Block Offset*
VBS	*VPLS endpoint Block Size*
VCI	*Virtual Channel Identifier*
VCL	*Virtual Channel Label*
VE-ID	*VPLS Endpoint Identifier*
VID	*VLAN Identification*
ViLTE	*Video over Long Term Evolution*
VLAN	*Virtual LAN*
VoLTE	*Voice over Long Term Evolution*)
V-PCRF	*Visited-PCRF*

VPI *Virtual Path Identifier*

VPLS *Virtual Private LAN Service*

VPN *Virtual Private Network*

VRF *VPN Routing and Forwarding*

VSI *Virtual Switch Instance*

W, X

WAN *Wide Area Network*

WB *Wide Band*

WDM *Wavelength Division Multiplexing*

WFQ *Weighted Fair Queueing*

WRED *Weighted Random Early Discard*

WRR *Weighted Round Robin*

XOR *eXclusive OR*

1

Mobile Network Architecture

1.1. Functional architecture

The functional architecture of the EPS (Evolved Packet System) network is described in Figure 1.1, when the mobile attaches to its home network.

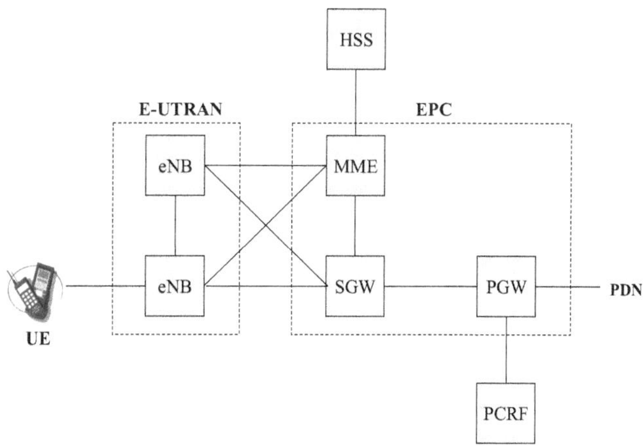

Figure 1.1. *Functional architecture of EPS network*

Two EPS network architectures are defined when the mobile is attached to a visited network, as part of roaming:

– HR (Home Routed) architecture, for which the PGW (PDN Gateway) entity is located in the home network (Figure 1.2);

– LBO (Local BreakOut) architecture, for which the PGW entity is located in the visited network (Figure 1.3).

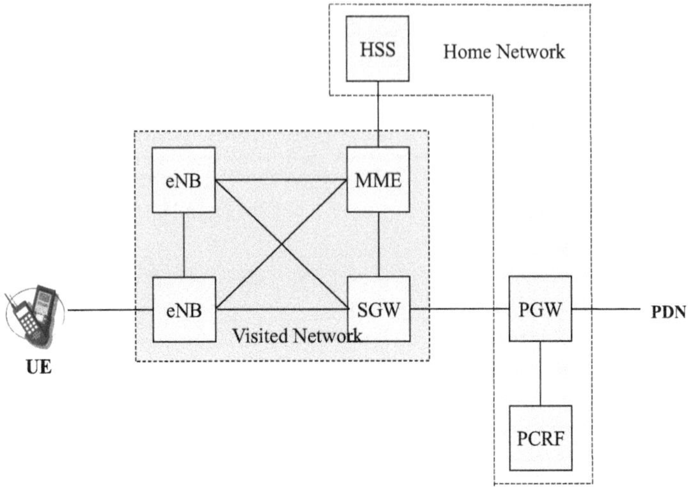

Figure 1.2. *HR architecture for EPS network*

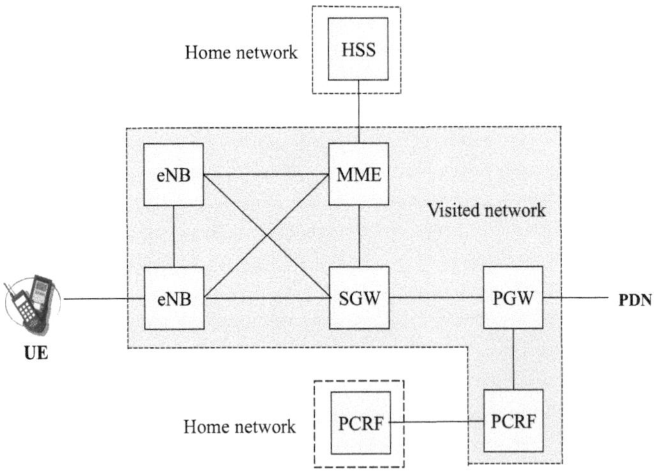

Figure 1.3. *LBO architecture for EPS network*

The EPS mobile network consists of an Evolved Packet Core (EPC) network and an Evolved Universal Terrestrial Radio Access Network (E-UTRAN).

The E-UTRAN access network ensures the connection of the User Equipment (UE) and the resource reservation on radio interface.

The EPC core network interconnects the access networks, provides the interface to the Packet Data Network (PDN) and ensures the attachment of mobiles, the authorization to access the service and the establishment of bearers.

1.1.1. *eNB entity*

The E-UTRAN access network includes a single type of entity, the evolved Node Base station (eNB), which connects to the mobiles.

The eNB entity is responsible for the management of radio resources, for the control of the establishment of the Data Radio Bearer (DRB), in which the mobile traffic is transmitted and for its mobility management during the session (handover).

The eNB entity transfers the traffic data from the mobile (respectively from the Serving Gateway (SGW) entity) to the SGW entity (respectively to the mobile).

When the eNB entity receives data from the mobile or the SGW entity, it refers to the QoS Class Identifier (QCI) for the implementation of the data scheduling mechanism.

The eNB entity can perform the marking of the DiffServ Code Point (DSCP) field of IP (Internet Protocol) header, based on the assigned QCI identifier, for outgoing data to the SGW entity.

The eNB entity performs compression and encryption of traffic data on the radio interface.

The eNB entity performs encryption and integrity control of signaling data exchanged with the mobile.

The eNB entity performs the selection of the Mobility Management Entity (MME), part of the core network, to which the mobile is attached.

The eNB entity treats paging requests sent by the MME entity for their distribution in the cell corresponding to the radio coverage area of the eNB entity.

The eNB entity also distributes system information to the cell, containing the technical characteristics of the radio interface, and allowing the mobile to connect.

The eNB entity uses the measurements made by the mobile to decide on the initiation of a cell change during a session (handover).

1.1.2. *MME entity*

The MME entity is the network control tower, allowing mobile access to the service and controlling bearer establishment for the transmission of traffic data.

The MME entities belong to a group (pool). Load balancing of the MME load is provided by the eNB entities within a group that must have access to each MME entity of the same group. The MME entity is responsible for attachment and detachment of the mobile.

During attachment, the MME entity retrieves the subscriber's profile and the subscriber's authentication data stored in the Home Subscriber Server (HSS) and performs authentication of the mobile.

During attachment, the MME entity registers the Tracking Area Identity (TAI) of the mobile and allocates a Globally Unique Temporary Identity (GUTI) to the mobile, which replaces the private International Mobile Subscriber Identity (IMSI).

The MME entity manages a list of location areas allocated to the mobile, where the mobile can move in an idle state, without contacting the MME entity to update its location area.

When attaching the mobile, the MME entity selects the SGW and PGW (PDN Gateway) entities for the construction of the default bearer, e.g. for the transport of IP packets containing SIP (Session Initiation Protocol) signaling or the data from Internet service.

For the construction of the bearer, the selection of the PGW entity is obtained from the Access Point Name (APN), communicated by the mobile or by the HSS entity in the subscriber's profile. The source MME entity also selects the target MME entity when the mobile changes both cell and group (pool).

The MME entity provides the information required for lawful interception, such as the mobile status (idle or connected), the location area if the mobile is idle or the E-UTRAN Cell Global Identifier (ECGI) if the mobile is in session.

1.1.3. *SGW entity*

The SGW entities are also organized into groups (pools). To ensure load balancing of SGW load, each eNB entity within a group must have access to each SGW entity of the same group.

The SGW entity forwards incoming data from the PGW entity to the eNB entity and outgoing data from the eNB entity to the PGW entity.

When the SGW entity receives data from the eNB or PGW entities, it refers to the QCI identifier for the implementation of the data scheduling mechanism.

The SGW entity can perform marking of the DSCP field of IP header based on the assigned QCI identifier for incoming and outgoing data.

The SGW entity is the anchor point for intra-system handover (mobility within EPS network) provided that the mobile does not change group. Otherwise, the PGW entity performs this function.

The SGW entity is also the anchor point at the inter-system handover in PS (Packet-Switched) mode, requiring the transfer of traffic data from the mobile to the 2nd or 3rd generation mobile network.

The SGW entity informs the MME entity of incoming data when the mobile is in idle state, which allows the MME entity to trigger paging of all eNB entities of the location area.

A mobile in the idle state remains attached to the MME entity. However, it is no longer connected to the eNB entity, and thus the radio bearer and the S1 bearer, built between the eNB and SGW entities, are deactivated.

In the case of roaming corresponding to the HR architecture, the SGW entity of the visited network diverts mobile traffic in the framework of lawful interceptions.

1.1.4. *PGW entity*

The PGW entity is the gateway router providing the EPS network connection to the PDN network (Internet network).

When the PGW entity receives data from the SGW entity or PDN network, it refers to the QCI identifier for the implementation of the data scheduling mechanism.

The PGW entity can perform DSCP marking of IP header based on the assigned QCI identifier.

During attachment, the PGW entity grants an IPv4 or IPv6 address to the mobile. If the assigned IPv4 address is a private address, the entity PGW performs Network Address and Port Translation (NAPT), of translating the IP address and the TCP or UDP port number.

The PGW entity constitutes the anchor point for inter-SGW mobility when the mobile changes groups.

The PGW entity hosts the Policy and Charging Enforcement Function (PCEF) which applies the rules relating to mobile traffic data on packet, of filtering, of charging and of Quality of Service (QoS) to be applied to the bearer to build.

The Policy Charging and Rules Function (PCRF) entity, outside the EPS network, provides the PCEF function of the PGW entity with the rules to apply when establishing bearers.

The PGW entity performs replication of the mobile traffic data within the framework of lawful interception, when the mobile is attached to its home network or to visited networks, for both types of architecture, HR or LBO.

1.1.5. *HSS entity*

The HSS entity is a database storing the data specific to each user. The main data stored include the identities of the users, the authentication parameters and the service profile.

When subscribing to the EPS network, the mobile is assigned a private identity (IMSI) which is associated with a service profile and a secret key (Ki).

The authentication parameters are used to control access to the mobile for attachment to the EPS network.

The service profile determines the services that mobile has subscribed.

1.1.6. *PCRF entity*

The PCRF entity provides to the PCEF entity, integrated in the PGW entity, the necessary information for the control and the charging of the traffic data (IP packets).

This information is stored in the Subscription Profile Repository (SPR) during the creation of the subscription.

Traffic control includes the following:

– association between a Service Data Flow (SDF) and EPS bearer;

– blocking or allowing IP packets;

– assignment of QCI parameter to EPS bearer.

The PCEF entity executes the rules provided by the PCRF entity to control the traffic flow, the accounting of traffic volume and the charging.

The PCEF entity may relate to the PCRF entity a change of state of a service flow, as in the case of loss of radio coverage of the mobile.

The PCRF entity may receive a session request from the AF (Application Function) entity as in the case of the establishment of a voice or conversational video communication initialized at the IMS (IP Multimedia Sub-system) network.

The PCRF entity may provide the AF entity information about events occurring in the mobile network as in the case of loss of radio coverage of the mobile.

In case of roaming, the PCEF entity located in the visited network requests rules to the Visited-PCRF (V-PCRF) entity, which gets them from the Home-PCRF (H-PCRF) entity.

1.2. Protocol architecture

The protocol architecture of the EPS network is described in Figure 1.4 for the control plane and in Figure 1.5 for the traffic plane.

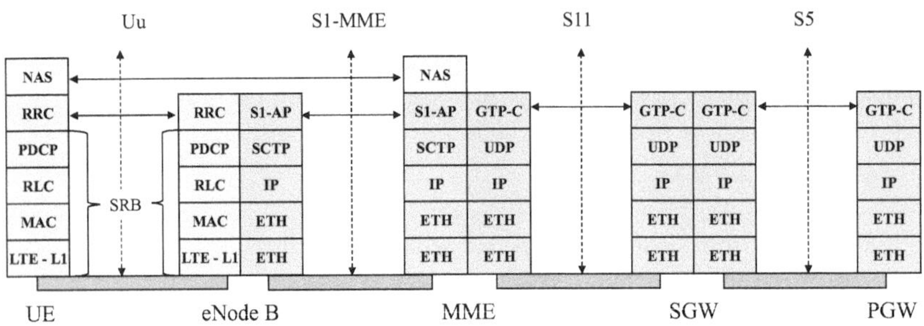

Figure 1.4. *Protocol architecture: control plane*

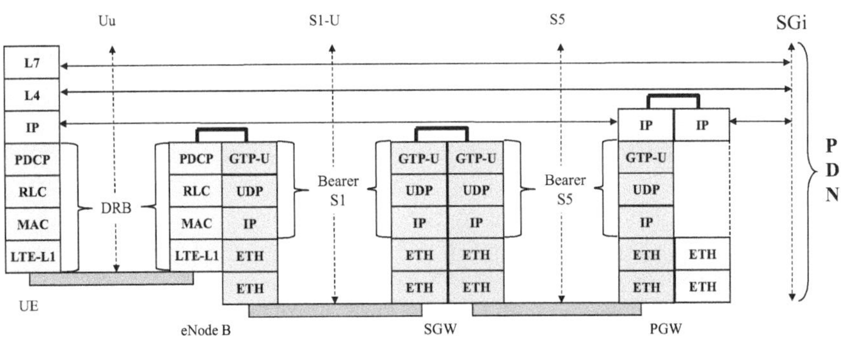

Figure 1.5. *Protocol architecture: traffic plane*

1.2.1. *Long-term evolution-Uu interface*

The LTE-Uu interface is the reference point between the mobile and the eNB entity.

This interface supports Radio Resource Control (RRC) signaling exchanged between the mobile and the eNB entity, transmitted in the Signaling Radio Bearer (SRB) and the mobile traffic data (IP packets) transmitted in the Data Radio Bearer (DRB).

The RRC protocol performs the following functions: broadcasting information system, control of the connection, control of the radio bearer, control of handover, transfer of measurement reports.

The RRC signaling also provides transport of the Non-Access Stratum (NAS) protocol exchanged between the mobile and the MME entity.

Traffic data corresponding to an IP packet and traffic data corresponding to a RRC message are encapsulated by the data link layer, which is divided into three sublayers:

– Packet Data Convergence Protocol (PDCP);

– Radio Link Control (RLC);

– Medium Access Control (MAC).

The PDCP protocol is used for RRC signaling messages related to dedicated control and for IP packet, and provides the following functions:

– compression of traffic data headers using the Robust Header Compression (ROHC) mechanism;

– security of traffic data (confidentiality) and of RRC signaling (integrity and confidentiality);

– delivery in sequence of RRC messages and IP packets;

– recovery of PDCP frames lost during the handover.

The RLC protocol provides control of the radio link between the mobile and the evolved Node B (eNB) entity.

The mobile can simultaneously activate multiple RLC instances, each instance corresponding to a Packet Data Convergence Protocol (PDCP) instance.

The RLC protocol operates in three operation modes:

– Acknowledged Mode (AM);

– Unacknowledged Mode (UM);

– Transparent Mode (TM) for which no header is added to data.

The RLC protocol does the following functions:

– retransmission in the case of error via the Automatic Repeat reQuest (ARQ) mechanism for the acknowledged mode only;

– concatenation, segmentation and reassembly of PDCP frames in both the acknowledged and unacknowledged mode;

– possible resegmentation of PDCP frames, in acknowledgment mode, during a retransmission of the RLC frame;

– resequencing of received data in both the acknowledged and unacknowledged mode;

– detection of duplicate data in both the acknowledged and unacknowledged mode;

The MAC protocol provides the following functions:

– multiplexing RLC frames from multiple instances in a transport block;

– allocation of the radio resources via a scheduling mechanism for both transmission directions;

– management of retransmission in the case of error via the Hybrid Automatic Repeat reQuest (HARQ) mechanism;

– management of the random access procedure.

The physical layer of LTE-Uu interface implements the following functions: error correction and error detection, multiplexing of physical channels, modulation and signal amplification.

1.2.2. *S1 interface*

The S1 interface is a group of interfaces between the E-UTRAN access network and the EPC core network and consists of the S1-MME and S1-U interface.

The S1-MME interface is the reference point between the MME and eNB entities for signaling. This interface supports S1-AP (Application Part) signaling.

The S1-AP protocol provides the following functions: establishment of the mobile context, establishment, modification and release of the EPS Radio Access Bearer (E-RAB), built between the mobile and the SGW entity, management of the handover, paging.

The S1-AP protocol also provides the transport of the NAS signaling exchanged between the mobile and the MME.

The S1-U interface is a reference point between the eNB and SGW entities. This interface supports GPRS Tunnel Protocol User (GTP-U), which encapsulates the IP packet flow.

1.2.3. *S11 interface*

The S11 interface is the reference point between the MME and SGW entities for signaling via the GPRS Tunnel Control Protocol (GTPv2-C).

The GTPv2-C protocol supports the following functions: managing the context of mobile and the S1 bearer, notification of incoming data when the mobile is on idle state.

1.2.4. *S5 and S8 interfaces*

The S5 interface is the reference point between the SGW and PGW entities for signaling via the GTPv2-C protocol and traffic data (IP packets) via the GTP-U protocol.

In case of roaming, for HR architecture, the S8 interface is the reference point between the SGW entity of the visited network and the PGW entity of the home network. This interface supports the same protocols as the S5 interface.

1.2.5. *S10 interface*

The S10 interface is the reference point between MME entities. This interface is used when the mobile changes group (pool) and the MME must be relocated. This interface supports GTPv2-C signaling.

1.2.6. *SGi interface*

The SGi interface is the reference point between the PDW entity and the PDN network (Internet).

1.2.7. *X2 interface*

The X2 interface is the reference point between two eNB entities. It is activated when both eNB entities belong to the same group.

The X2 interface supports the X2-AP signaling for the control plane (Figure 1.6), and GTP-U protocol for the IP packet flow when the mobile changes cell (Figure 1.7).

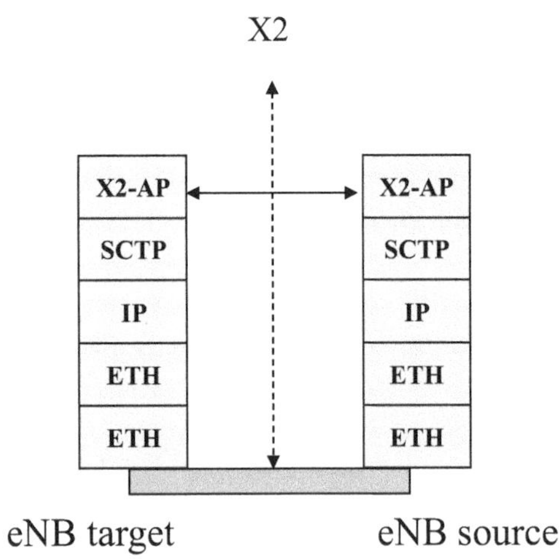

Figure 1.6. *Protocol architecture of the X2 interface: control plane*

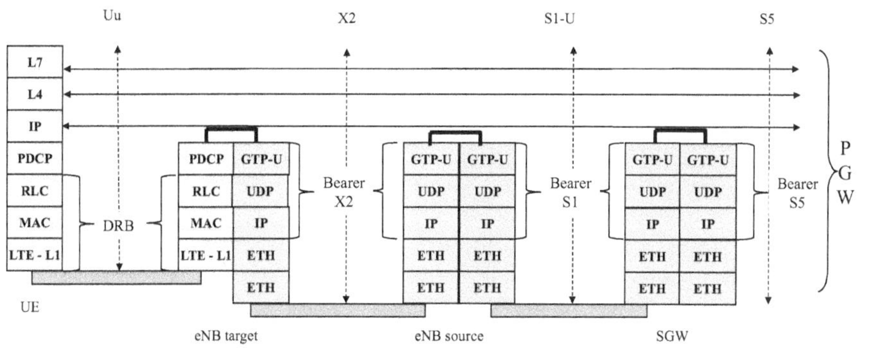

Figure 1.7. *Protocol architecture: traffic plane during handover based on the X2 interface*

The tunnel established between the two eNB entities is unidirectional (eNB source to target eNB). It allows us to transfer to the target eNB entity the received traffic data of SGW entity. It is established temporarily during the handover of the mobile.

1.2.8. *S6a interface*

The S6a interface is the reference point between MME and HSS entities. This interface supports the DIAMETER signaling.

The DIAMETER protocol allows the MME to recover the authentication data and the service profile of the mobile.

In case of roaming, for HR and LBO architectures, the S6a interface is the reference point between the MME entity of the visited network and the HSS entity of the home network.

1.2.9. *Gx interface*

The Gx interface is the reference point between the PCRF entity and the PCEF function of the PGW entity. This interface supports the DIAMETER signaling.

The DIAMETER protocol allows the PGW entity to recover rules applying to the EPS bearer and to inform the PCRF entity to the termination of the session on the EPS network.

1.2.10. *Rx interface*

The Rx interface is the reference point between the PCRF and AF entities. This interface supports the DIAMETER signaling.

The DIAMETER protocol allows the AF entity:

– to provide the characteristics of the media so that the EPS network can establish the dedicated support;

– to inform the session is terminated, so the EPS network can release the dedicated support.

1.2.11. *S9 interface*

The S9 interface is the reference point between the H-PCRF entity located in the home network and the V-PCRF entity located in the visited network. This interface supports the DIAMETER signaling.

2

Mobile Network Signaling Protocols

2.1. NAS protocol

The Non-Access Stratum (NAS) protocol is the signaling exchanged between the User Equipment (UE) and the Mobility Management Entity (MME).

The NAS protocol is transported by the RRC (Radio Resource Control) protocol over the radio interface LTE-Uu and by the S1-AP (Application Part) protocol over the S1-MME interface.

The NAS protocol comprises the following two protocols:

– EMM (EPS Mobility Management). This takes care of controlling mobility and security.

– ESM (EPS Session Management). This controls the bearer establishment.

From the point of view of the MME entity, the mobile can be in one of two operational states: either attached (EMM-REGISTERED) or detached (EMM-DEREGISTERED).

In the EMM-DEREGISTERED state, the mobile location is not known to the MME entity, and, therefore, it cannot be contacted.

The switch to the registered state takes place when the mobile attaches, which comprises the following four procedures:

– mutual authentication of the mobile and the MME entity;

– registration of the mobile location with the MME entity;

– assignment of the Globally Unique Temporary Identity (GUTI) to the mobile;

– establishment of the default bearer.

The switch to the deregistered state takes place when the mobile detaches or when the attachment of the mobile, the update of its location or the service request are rejected by the MME entity.

2.1.1. *EMM messages*

2.1.1.1. *Attachment and detachment*

The procedure of attachment is initiated by the mobile in the deregistered state, by sending the message EMM ATTACH REQUEST to the MME entity.

This message contains the mobile identity, GUTI or IMSI (International Mobile Subscriber Identity), and the Tracking Area Identity (TAI).

The mobile can attach the message ESM PDN CONNECTIVITY REQUEST to establish the default bearer.

Upon receiving this message, the MME entity begins the authentication and security procedures for the NAS protocol.

If they are successfully completed, the MME entity responds with the message EMM ATTACH ACCEPT, containing a new GUTI, and the message ESM ACTIVATE DEFAULT EPS BEARER CONTEXT REQUEST, to establish the default bearer.

The mobile responds with the message EMM ATTACH COMPLETE containing the message ESM ACTIVATE DEFAULT EPS BEARER CONTEXT ACCEPT, to acknowledge the previous message.

If the procedures are not successful, the MME entity responds with the message EMM ATTACH REJECT, containing the message ESM PDN CONNECTIVITY REJECT, which causes the mobile to detach.

Detachment may be initiated by the mobile or the MME entity by sending the message EMM DETACH REQUEST.

The response EMM DETACH ACCEPT concludes the detachment procedure. The response is not transmitted by the MME entity if the detach request sent by the mobile indicates that it has been turned off. The detachment procedure implicitly causes the release of the active bearers.

2.1.1.2. *Authentication*

The procedure of mutual authentication is initiated by the MME entity by sending the message EMM AUTHENTICATION REQUEST, containing a random number RAND and the Authentication Network (AUTN).

The mobile uses the RAND received to locally compute its own authentication code RES (Result) and that of the network (AUTN), and compares the AUTN calculated to the one received from the MME entity.

If the MME entity is authenticated, the mobile responds with the message EMM AUTHENTICATION RESPONSE, containing the authentication code RES.

Otherwise, it indicates that network authentication failed with the message EMM AUTHENTICATION FAILURE.

The MME entity compares the RES value received from the mobile with that communicated by the Home Subscriber Server (HSS).

If the two codes are the same, the mobile is authenticated and the MME entity triggers security mode for NAS signaling.

Otherwise, the MME entity responds with the message EMM AUTHENTICATION REJECT.

2.1.1.3. *Security mode*

When mutual authentication has been successful, the MME begins putting the NAS signaling in security mode by sending the message EMM SECURITY MODE COMMAND. The integrity of this message is protected.

If the check on the integrity of the message EMM SECURITY MODE COMMAND is positive, the mobile responds with the message EMM SECURITY MODE COMPLETE. All subsequent NAS messages are encrypted and their integrity is checked.

If the check on the integrity of the message EMM SECURITY MODE COMMAND is negative, the mobile responds with the message EMM SECURITY MODE REJECT.

2.1.1.4. *Tracking area update*

The procedure of tracking area update is periodically initiated so that the mobile can maintain its tracking area, or when the mobile has changed its location area. The mobile, in the registered state, sends the message EMM TRACKING AREA UPDATE REQUEST to the MME entity.

The MME entity responds either with the message EMM TRACKING AREA UPDATE ACCEPT if it accepts the update, or else with the message EMM TRACKING AREA UPDATE REJECT, indicating the cause of the rejection.

If the message EMM TRACKING AREA UPDATE ACCEPT attributes a new GUTI to the mobile, the mobile confirms receipt of this by sending the message EMM TRACKING AREA UPDATE COMPLETE.

2.1.1.5. *Service request*

The service request is sent, when the mobile is in idle mode, to re-establish the default bearers on the LTE-Uu and S1-U interfaces.

The service request is initiated by the mobile, by sending the EMM SERVICE REQUEST when signaling or traffic data is waiting.

The mobile is notified of awaiting data at the level of the network by way of the paging procedure.

The MME entity may reject the request, in which case it responds with the message EMM SERVICE REJECT. This response causes the switch of the mobile to the deregistered state.

2.1.2. *ESM messages*

The mobile sends the request to establish the default bearer when the mobile attaches to the MME entity.

The establishment request for the dedicated bearer can be transmitted by the network or the mobile.

The dedicated bearer corresponds to a specific request by the mobile, following for example the establishment of a voice or a conversational video.

The dedicated bearer is associated with a particular quality of service, corresponding to a particular QoS Class Identifier (QCI) which is different to that of the default bearer.

Table 2.1 summarizes the ESM messages exchanged for the establishment, modification and release of the default and dedicated bearers.

Establishment of default bearer, initiated by the network		
Source	Message	Destination
MME	ACTIVATE DEFAULT EPS BEARER CONTEXT REQUEST	UE
UE	ACTIVATE DEFAULT EPS BEARER CONTEXT ACCEPT or ACTIVATE DEFAULT EPS BEARER CONTEXT REJECT	MME

Establishment of default bearer, initiated by the mobile		
Source	Message	Destination
UE	PDN CONNECTIVITY REQUEST	MME
MME	ACTIVATE DEFAULT EPS BEARER CONTEXT REQUEST or PDN CONNECTIVITY REJECT	UE

Establishment of dedicated bearer, initiated by the network		
Source	Message	Destination
MME	ACTIVATE DEDICATED EPS BEARER CONTEXT REQUEST	UE
UE	ACTIVATE DEDICATED EPS BEARER CONTEXT ACCEPT or ACTIVATE DEDICATED EPS BEARER CONTEXT REJECT	MME

Modification of dedicated bearer, initiated by the network		
Source	Message	Destination
MME	MODIFY EPS BEARER CONTEXT REQUEST	UE
UE	MODIFY EPS BEARER CONTEXT ACCEPT or MODIFY EPS BEARER CONTEXT REJECT	MME

Release of dedicated bearer, initiated by the network		
Source	Message	Destination
MME	DEACTIVATE EPS BEARER CONTEXT REQUEST	UE
UE	DEACTIVATE EPS BEARER CONTEXT ACCEPT	MME

Establishment of dedicated bearer, initiated by the mobile		
Source	Message	Destination
UE	BEARER RESOURCE ALLOCATION REQUEST	MME
MME	ACTIVATE DEDICATED EPS BEARER CONTEXT REQUEST or ACTIVATE DEDICATED EPS BEARER CONTEXT REJECT or MODIFY EPS BEARER CONTEXT REQUEST	UE

Modification of dedicated bearer, initiated by the mobile		
Source	Message	Destination
UE	BEARER RESOURCE MODIFICATION REQUEST	MME
MME	ACTIVATE DEDICATED EPS BEARER CONTEXT REQUEST or MODIFY EPS BEARER CONTEXT REQUEST or DEACTIVATE EPS BEARER CONTEXT REQUEST or BEARER RESOURCE MODIFICATION REJECT	UE

Release of dedicated bearer, initiated by the mobile		
Source	Message	Destination
UE	PDN DISCONNECT REQUEST	MME
MME	DEACTIVATE EPS BEARER CONTEXT REQUEST or PDN DISCONNECT REJECT	UE

Table 2.1. *ESM messages*

2.2. RRC protocol

The RRC protocol is the signaling exchanged between the mobile and the evolved Node Base station (eNB) over the LTE-Uu radio interface.

The RRC protocol performs the following functions:

– Broadcast of system information related to the characteristics of the radio interface.

– Control of the RRC connection. This procedure includes the paging, establishment, modification and release of the Signaling Radio Bearer (SRB) and the Data Radio Bearer (DRB). It also includes the activation of security mode over the LTE-Uu interface, the procedure which consists of putting mechanisms in place to encrypt the traffic and RRC signaling flows and to control the integrity of the RRC signaling flows.

– Control of handover. This procedure executes the changing of cell between two eNB entities (intrasystem handover) or between an eNB entity and a base station from a 2nd and 3rd-generation mobile network (intersystem handover).

– Measurement reporting. The eNB entity can trigger measurements carried out by the mobile, either periodically or on demand, to prepare for handover.

– Transport of the NAS messages exchanged between the mobile and the MME entity.

From the point of view of the eNB entity, the mobile may be in one of two operational states: idle mode (RRC_IDLE) or connected mode (RRC_CONNECTED).

In idle mode, the mobile is not known to the eNB entity. It remains in this state until the RRC connection setup procedure is completed. The setup procedure is triggered by the mobile when it wishes to transmit traffic or signaling data. In that state, the mobile used the SRB0 bearer.

In connected mode, the mobile can transmit and receive signaling and traffic data. The mobile is attributed to an identifier that is unique to the cell, the Cell Radio Network Temporary Identity (C-RNTI). In this state, the mobile uses either the SRB1 bearer for RRC messages with possible associated NAS messages, or the SRB2 bearer for RRC messages transporting solely NAS messages.

Table 2.2 summarizes the type of SRB, the mode of RLC (Radio Link Control) protocol and the channels used by the different RRC messages over the radio interface.

MasterInformationBlock				
SRB	RLC mode	Logical channel	Transport channel	Physical channel
Not available	TM	BCCH	BCH	PBCH

SystemInformationBlock				
SRB	RLC mode	Logical channel	Transport channel	Physical channel
Not available	TM	BCCH	DL-SCH	PDSCH

Paging				
SRB	RLC mode	Logical channel	Transport channel	Physical channel
Not available	TM	PCCH	PCH	PDSCH

ConnectionSetup *ConnectionReject* *ConnectionReestablishment* *ConnectionReestablishmentReject*				
SRB	RLC mode	Logical channel	Transport channel	Physical channel
SRB0	TM	CCCH	DL-SCH	PDSCH

ConnectionRequest *ConnectionReestablishmentRequest*				
SRB	RLC mode	Logical channel	Transport channel	Physical channel
SRB0	TM	CCCH	UL-SCH	PUSCH

ConnectionReconfiguration *ConnectionRelease* *SecurityModeCommand*				
SRB	RLC mode	Logical channel	Transport channel	Physical channel
SRB1	AM	DCCH	DL-SCH	PDSCH

DLInformationTransfer (1)				
SRB	RLC mode	Logical channel	Transport channel	Physical channel
SRB2	AM	DCCH	DL-SCH	PDSCH

(1): transport of NAS messages only, downstream

ULInformationTransfer (2)				
SRB	RLC mode	Logical channel	Transport channel	Physical channel
SRB2	AM	DCCH	UL-SCH	PUSCH

(2): transport of NAS messages only, upstream

ConnectionSetupComplete *SecurityModeComplete* *SecurityModeFailure* *ConnectionReconfigurationComplete* *ConnectionReestablishmentComplete* *MeasurementReport*				
SRB	RLC mode	Logical channel	Transport channel	Physical channel
SRB1	AM	DCCH	UL-SCH	PUSCH

Table 2.2. *RRC messages*

2.2.1. *System information*

The information relating to the radio interface is divided between the messages *MasterInformationBlock* and *SystemInformationBlock*.

These messages are transmitted periodically, and a change in these data is notified to the mobile by paging.

The *MasterInformationBlock* message contains the following data:

– the bandwidth of the radio signal for the downstream direction (1.4, 3, 5, 10, 15, 20 MHz);

– the System Frame Number (SFN);

– the configuration of the physical channel PHICH of the radio interface. The configuration of this physical channel is defined by the operator.

All *SystemInformationBlock* messages, with the exception of the *SystemInformationBlockType1* message, are mapped in a *SystemInformation* message.

Each *SystemInformation* message contains *SystemInformationBlock* messages with the same periodicity.

The *SystemInformationBlockType2* message must necessarily be mapped in the *SystemInformation1* message.

The *SystemInformationBlockType1* message contains the following data:

– the Mobile Country Code (MCC) and Mobile Network Code (MNC) of the mobile network;

– the Tracking Area Code (TAC) of the location area. The identity of the TAI is a concatenation of the MCC, MNC and TAC codes;

– the E-UTRAN Cell Global Identifier (ECGI);

– the periodicity of the *SystemInformation* messages and the types of *SystemInformationBlock* messages that they contain.

Table 2.3 shows the data transported by the different types of *SystemInformationBlock* message.

2.2.2. *Control of RRC connection*

SystemInformationBlockType2	Bandwidth in upstream direction Configuration of physical channels
SystemInformationBlockType3	Cell selection parameters
SystemInformationBlockType4	EPS neighboring cells, same frequency
SystemInformationBlockType5	EPS neighboring cells, different frequency
SystemInformationBlockType6	UMTS neighboring cells
SystemInformationBlockType7	GSM/GPRS neighboring cells
SystemInformationBlockType8	CDMA 2000 neighboring cells
SystemInformationBlockType9	Identity of the Home eNB (HeNB)
SystemInformationBlockType10 SystemInformationBlockType11	Earthquake and Tsunami Warning System (ETWS)
SystemInformationBlockType12	Commercial Mobile Alert System (CMAS)
SystemInformationBlockType13	Information related to the MBSFN (MBMS over Single Frequency Network) area

Table 2.3. SystemInformationBlock *messages*

The different procedures associated with the control of the RRC connection relate to paging, RRC connection setup, activation of security mode, RRC connection reconfiguration, RRC connection re-establishment and RRC connection release.

The *Paging* message is used by the eNB entity to alert one or more mobiles in the RRC_IDLE state.

The *Paging* message also helps to inform the mobile in RRC_IDLE or RRC_CONNECTED state about a change in the system information or about a notification on ETWS transmitted in the *SystemInformationBlockType10* and

SystemInformationBlockType11 messages or CMAS transmitted in the *SystemInformationBlockType12* message.

The RRC *ConnectionRequest* message is used by the mobile to request the establishment of an RRC connection.

The RRC *ConnectionSetup* message is used by the eNB entity to configure the SRB1 bearer.

The RRC *ConnectionSetupComplete* message is used by the mobile to confirm the setup of the RRC connection. This message can also transport NAS messages.

The RRC *ConnectionReject* message is used by the eNB entity to reject the request for the RRC connection.

Upon receiving the context about the mobile from the MME entity, the eNB entity activates security mode for the radio interface, by sending the *SecurityModeCommand* message.

The *SecurityModeCommand* message is only checked for integrity.

The *SecurityModeComplete* message is used by the mobile to confirm the activation of security mode.

The *SecurityModeFailure* message is used by the mobile to indicate that security mode was unable to be activated.

The encryption of the RRC messages will be effective only if the procedure has been successful.

Having initiated the security mode activation procedure, the eNB entity begins the activation of the DRB. The RRC messages are encrypted and checked for integrity.

The RRC *ConnectionReconfiguration* message is used by the eNB entity to command a modification of the RRC connection. This message may relate to the configuration of the measurements, control of the mobility and configuration of the DRB default bearer. This message can also transport NAS messages.

The message RRC *ConnectionReconfigurationComplete* message is used by the mobile to confirm the reconfiguration of the RRC connection.

The RRC *ConnectionReestablishmentRequest* message is used by the mobile to request the re-establishment of the RRC connection.

The RRC *ConnectionReestablishment* message is used by the eNB entity to re-establish the RRC connection.

The RRC *ConnectionReestablishmentComplete* message is used by the mobile to confirm the re-establishment of the RRC connection.

The RRC *ConnectionReestablishmentReject* message is used by the eNB entity to indicate that the re-establishment of the RRC connection has been rejected.

The RRC *ConnectionRelease* message is used by the eNB entity to release the RRC connection. The procedure can also be used to redirect the mobile to a different frequency band.

2.2.3. *Measurement report*

The measurements carried out by the mobile on the serving cell and the neigboring cells must be in line with the configuration indicated in the RRC *ConnectionReconfiguration* message transmitted by the eNB entity.

The mobile sends the eNB entity the measurements in the RRC *MeasurementReport* message.

Measurements carried out on the serving cell and the neighboring cells are used for the selection of the cell and for the handover.

Intra-frequency measurement, essential for mobility, is configured when connecting.

Inter-frequency and inter-RAT (Radio Access Technology) measurement can also be configured when connecting.

As the mobile does not generally have several radio receivers, the inter-frequency and inter-RAT measurement should be performed at intervals arranged in the frame.

The configuration of measurements to be performed by the mobile is triggered by the eNB entity in RRC *ConnectionSetup, ConnectionReconfiguration, ConnectionReestablishment* messages.

The measurement configurations to perform define the following parameters:

– the object that identifies the radio channel;

– the event triggering the measurement report;

– the combination of objects and events;

– the measurement filtering parameter;

– the periodicity of measurements.

2.3. S1-AP protocol

The S1-AP protocol is the signaling exchanged between the eNB and MME entities over the S1-MME interface.

The S1-AP protocol performs the following functions:

– activation of the context of the mobile by the MME entity;

– establishment, modification and release of the EPS Radio Access Bearer (E-RAB);

– handover management;

– paging. This procedure tells the eNB entity that the message needs to be broadcast in the cell;

– transport of the NAS signaling exchanged between the mobile and the MME entity;

– establishment of the S1-MME interface.

Table 2.4 summarizes the S1-AP messages sent to paging, context management, bearer management, mobility management, management of the S1-MME interface and NAS messages transport.

Function	Request	Response
Context management	INITIAL CONTEXT SETUP REQUEST	INITIAL CONTEXT SETUP RESPONSE or INITIAL CONTEXT SETUP FAILURE
	UE CONTEXT RELEASE COMMAND	UE CONTEXT RELEASE COMPLETE

Function	Request	Response
Bearer management	E-RAB SETUP/MODIFY REQUEST	E-RAB SETUP/MODIFY RESPONSE
	E-RAB RELEASE COMMAND	E-RAB RELEASE RESPONSE
	E-RAB RELEASE INDICATION	None

Function	Request	Response
Paging	PAGING	None

Function	Request	Response
Mobility management	HANDOVER REQUIRED	HANDOVER COMMAND
	HANDOVER REQUEST	HANDOVER REQUEST ACKNOWLEDGE or HANDOVER FAILURE
	eNB STATUS TRANSFER	None
	MME STATUS TRANSFER	None
	HANDOVER NOTIFY	None
	PATH SWITCH REQUEST	PATH SWITCH ACKNOWLEDGE or PATH SWITCH FAILURE

Function	Request	Response
Management of the S1-MME interface	S1 SETUP REQUEST	S1 SETUP RESPONSE or S1 SETUP FAILURE
	ENB CONFIGURATION UPDATE	ENB CONFIGURATION UPDATE ACKNOWLEDGE or ENB CONFIGURATION UPDATE FAILURE
	MME CONFIGURATION UPDATE	MME CONFIGURATION UPDATE ACKNOWLEDGE or ENB CONFIGURATION UPDATE FAILURE
	OVERLOAD START	None
	OVERLOAD STOP	None
Transport of NAS signaling	INITIAL UE MESSAGE	None
	DOWNLINK NAS TRANSPORT	None
	UPLINK NAS TRANSPORT	None

Table 2.4. *S1-AP messages*

2.3.1. *Context management*

The context of the mobile has to be established at the level of the eNB and MME entities so as to transmit the mobile traffic and the NAS signaling.

The context of the mobile includes the contexts relating to the default bearer, security, the capacities of the mobile and roaming restrictions.

Context setup for the mobile begins with the INITIAL CONTEXT SETUP REQUEST message transmitted by the MME entity to the eNB entity. This message follows the reception of the INITIAL UE MESSAGE message.

If the eNB is unable to establish the context of the mobile, it responds with the INITIAL CONTEXT SETUP FAILURE message.

Release of the context of the mobile is done by way of the UE CONTEXT RELEASE COMMAND message transmitted by the MME entity to the eNB entity, for instance when the mobile changes cell.

This message is acknowledged in return by the UE CONTEXT RELEASE COMPLETE response.

2.3.2. *Bearer management*

Establishment and modification of the E-RAB dedicated bearer are initiated by the MME entity by sending the E-RAB SETUP/MODIFY REQUEST messages.

The eNB entity responds positively or negatively by sending the E-RAB SETUP/MODIFY RESPONSE messages.

Release of the dedicated bearer is initiated by the MME entity by sending the E-RAB RELEASE COMMAND message or by the eNB entity by sending an E-RAB RELEASE INDICATION message. Upon receiving this message, the MME entity begins the procedure of release of the dedicated bearer.

2.3.3. *Mobility management*

The decision regarding handover based on the S1 interface is made by the source eNB entity.

The phase of handover preparation begins with the sending of the HANDOVER REQUIRED message to the MME entity.

When the reservation of resources by the target eNB entity is effective, the MME entity responds with the HANDOVER COMMAND message.

The MME entity asks the target eNB entity to reserve the radio resources by way of the HANDOVER REQUEST message.

If the operation is successful, the target eNB entity responds with the HANDOVER REQUEST ACKNOWLEDGE message. This message can contain the elements for construction of a GTP-U tunnel to transfer the received data from the source eNB entity to the target eNB entity so they can be transmitted to the mobile.

If not, the target eNB entity responds with the HANDOVER FAILURE message.

The source eNB entity has to transfer the value of the field SN (Sequence Number) of the Packet Data Convergence Protocol (PDCP) to the target eNB entity in order to preserve the continuity of the PDCP frame numbering. This operation is done by the transmission of the following messages:

– eNB STATUS TRANSFER from the source eNB entity to the MME entity;

– MME STATUS TRANSFER from the MME entity to the target eNB entity.

This procedure applies only to bearers who use the Acknowledgment Mode (AM) of the RLC protocol, which is not the case of dedicated bearer for voice or video.

When the execution of the handover has been completed, the target eNB entity advises the MME entity of this by way of the HANDOVER NOTIFY message.

Regarding handover based on the X2 interface, the PATH SWITCH REQUEST message is transmitted by the target eNB entity to the MME entity for the transfer of the extremity of the GTP-U tunnel of the source eNB entity to the target eNB entity.

The MME responds with the PATH SWITCH ACKNOWLEDGE message if the response is positive or with PATH SWITCH FAILURE message if not.

2.3.4. *S1-MME interface management*

The eNB entity takes the initiative to activate the S1-MME interface by transmitting the S1 SETUP REQUEST message, indicating the list of serving location area.

The S1 SETUP RESPONSE message from MME entity contains information relating to the MME entity such as its code number, the pool number to which it belongs and the MNC and MCC codes.

The MME entity may respond negatively with the S1 SETUP FAILURE message.

Updates to the information about the eNB entity (or, respectively, the MME entity) are transmitted by the ENB CONFIGURATION UPDATE message (or, respectively, MME CONFIGURATION UPDATE message).

The sender of these messages receives a positive response with the ENB/MME CONFIGURATION UPDATE ACKNOWLEDGE messages or a negative one with the ENB/MME CONFIGURATION UPDATE FAILURE messages.

The MME entity notifies the eNB entity of the beginning (or, respectively, the end) of a state of overload by the OVERLOAD START message (or, respectively, OVERLOAD STOP message) so as to avoid being selected for the attachment of a new mobile.

2.4. X2-AP protocol

The X2-AP protocol is the signaling exchanged between two eNB entities over the X2 interface. The X2-AP protocol performs the following functions:

– Mobility management. This function enables the source eNB entity to transfer the connection of a mobile to the target eNB entity.

– Load management. This function is used by the eNB entities to provide an indication of the load of the cells that they serve.

– X2 interface management. This function is used for the activation of the X2 interface, the reconfiguration and reinitialization of the X2 interface.

Table 2.5 summarizes the X2-AP messages exchanged for mobility management, load management and X2 interface management.

Function	Request	Response
Mobility management	HANDOVER REQUEST	HANDOVER REQUEST ACKNOWLEDGE or HANDOVER PREPARATION FAILURE
	SN STATUS TRANSFER	None
	UE CONTEXT RELEASE	None
	HANDOVER CANCEL	None
Function	Request	Response
Load management	LOAD INFORMATION	None
	RESOURCE STATUS REQUEST	RESOURCE STATUS RESPONSE or RESOURCE STATUS FAILURE
	RESOURCE STATUS UPDATE	None
Function	Request	Response
X2 interface management	X2 SETUP REQUEST	X2 SETUP RESPONSE or X2 SETUP FAILURE
	ENB CONFIGURATION UPDATE	ENB CONFIGURATION UPDATE ACKNOWLEDGE or ENB CONFIGURATION UPDATE FAILURE
	RESET REQUEST	RESET RESPONSE

Table 2.5. *X2-AP messages*

2.4.1. *Mobility management*

The function of mobility management contains the following elementary procedures:

– handover preparation;

– transfer of the state of the SN field of the PDCP protocol;

– deactivation of the context of the mobile;

– handover cancellation.

The procedure of handover preparation is initiated by the source eNB entity by transmission of the HANDOVER REQUEST message to the target eNB entity.

The target eNB reserves the resources and responds with the HANDOVER REQUEST ACKNOWLEDGE message.

This message contains the value of the Tunnel Endpoint Identifier (TEID) used by the GTP-U protocol for the traffic transferred by the source eNB entity to the target eNB entity.

If the resources are unavailable, the target eNB entity sends back the HANDOVER PREPARATION FAILURE message.

The procedure of transfer of the state of the SN field consists of transferring to the eNB entity the value of the Sequence Number of the PDCP protocol with the SN STATUS TRANSFER message. At the source eNB entity, this message stops the attribution of the SN of the PDCP protocol for the downstream direction.

The procedure for context release of the mobile is initiated by the target eNB entity by sending the UE CONTEXT RELEASE message to the source eNB entity. Upon receiving this message, the source eNB entity eliminates the context of the mobile.

The procedure for cancelling the handover is initiated by the source eNB entity with the HANDOVER CANCEL message. This message causes the target eNB entity to release the resources on the radio interface.

2.4.2. *Load management*

The function of load management includes the following elementary procedures:

– cell load indication;

– initialization of resource status reports;

– resource status reporting.

The procedure for indication of the load of the cell is initiated by either of the eNB entities with the LOAD INFORMATION message. This message may contain the following elements of information:

– Interference Overload Indication (IOI). This information relates to the interference detected by the eNB entity for the upstream direction. The eNB entity receiving this information has to decrease the interference transmitted by the mobile;

– High Interference Indication (HII). This information relates to the interference detected by the eNB entity, for the upstream direction, indicating which bandwidths are affected. The eNB entity receiving this information needs to avoid using the said bandwidth, for the upstream direction, for the mobiles located on the periphery of the cell;

– Relative Narrowband Tx Power (RNTP). This information relates to a decrease in the power transmitted by an eNB entity. The eNB entity receiving this information includes it in its traffic management mechanism.

The procedure for initialization of resource status reporting is initiated by either of the eNB entities with the RESOURCE STATUS REQUEST message.

The eNB receiving this message responds with the RESOURCE STATUS RESPONSE message, which may contain status information for the radio resources, the S1 interface and the load of the eNB entity.

The eNB entity may respond with the RESOURCE STATUS FAILURE message if the reports cannot be generated.

The resource status report is then transmitted periodically by the eNB entity by sending the RESOURCE STATUS UPDATE message.

2.4.3. *X2 interface management*

The X2 interface is set up with the intention of exchanging the configuration data necessary for both eNB entities to function correctly.

One of the eNB entities initiates the procedure by indication of the cells served in the X2 SETUP REQUEST message to a candidate eNB entity.

The candidate eNB entity responds with the X2 SETUP RESPONSE message, also containing the list of serving cells.

The information communicated may also include the list of neighboring cells and the number of antennas for each serving cell.

The eNB entity may refuse the establishment of the X2 interface by sending the X2 SETUP FAILURE message in response.

The X2 interface setup is followed by configuration updating of the eNB entity if the configuration of the eNB entity changes. The configuration update is initiated by the ENB CONFIGURATION UPDATE message.

The remote eNB entity may respond positively with the CONFIGURATION UPDATE ACKNOWLEDGE message or negatively with the ENB CONFIGURATION UPDATE FAILURE message.

The reset of the X2 interface is intended to align the resources of the eNB entities in the case of an unexpected breakdown. The procedure is initiated by the RESET REQUEST message.

The receiving eNB entity responds with the RESET RESPONSE message. The procedure does not affect the data exchanged during the X2 setup or the configuration update of the eNB entity.

2.5. GTPv2-C protocol

GTP-U (GPRS Tunnel Protocol User) tunnels are used between two entities of the EPS network. GTP-U traffic tunnels are constructed on the S1-U, S5 and X2 interfaces.

The GTP-U tunnel is identified by the TEID parameter, the IP addresses and the UDP port numbers. The entity receiving the traffic or signaling data determines the value of the TEID parameter which the sending entity has to use.

The values of the TEID parameter of the GTP-U protocol are exchanged via the GTPv2-C (GPRS Tunnel Protocol Control), S1-AP and X2-AP protocols.

The TEID parameter used for the signaling exchanged over the S5 interface is unique. The same parameter is used for all signaling messages relating to the activation of the various S5 bearers for the different mobiles.

The TEID parameter used for the signaling exchanged over the S10 and S11 interfaces is unique for each mobile. The same parameter is used for all signaling messages relating to the establishment of the various S1-U bearers for the same mobile.

Table 2.6 summarizes the GTPv2-C messages exchanged for the management of support and mobility.

Type of messages	Request	Response
Bearer management	CREATE/DELETE SESSION REQUEST	CREATE/DELETE SESSION RESPONSE
	CREATE/MODIFY/DELETE BEARER REQUEST	CREATE/MODIFY/DELETE BEARER RESPONSE
	DOWNLINK DATA NOTIFICATION	DOWNLINK DATA NOTIFICATION ACKNOWLEDGE or DOWNLINK DATA NOTIFICATION FAILURE INDICATION
	CREATE/DELETE INDIRECT DATA FORWARDING TUNNEL REQUEST	CREATE/DELETE INDIRECT DATA FORWARDING TUNNEL RESPONSE

Type of messages	Request	Response
Mobility management	FORWARD RELOCATION REQUEST	FORWARD RELOCATION RESPONSE
	FORWARD RELOCATION NOTIFICATION	FORWARD RELOCATION ACKNOWLEDGE
	FORWARD ACCESS CONTEXT NOTIFICATION	FORWARD ACCESS CONTEXT ACKNOWLEDGE
	CONTEXT REQUEST	CONTEXT RESPONSE
	CONTEXT ACKNOWLEDGE	

Table 2.6. *GTPv2-C messages*

2.5.1. *Bearer management*

The signaling bearer related to a mobile is created by the CREATE SESSION REQUEST message. It is reinforced by the use of a TEID parameter. The message is transmitted:

– by the MME entity to the Serving Gateway (SGW) over the S11 interface;

– by the target SGW entity for the PDN Gateway (PGW) over the S5 interface.

The request is transmitted when any of the following procedures are initiated:

– attachment of the mobile;

– traffic request from the mobile;

– updating of the Tracking Area Code (TAC);

– handover.

The SGW entity (or, respectively, PGW entity) responds to the MME entity (or, respectively, SGW entity) with the CREATE SESSION RESPONSE message.

The signaling bearer is deactivated by the exchange of the DELETE SESSION REQUEST/RESPONSE messages.

The procedure is triggered when the mobile detaches, when the traffic is released, when the TAC changes, leading to a modification of the SGW entity, or when handover occurs, with a switch of the SGW.

Similarly, the dedicated bearer specific to a mobile is created, possibly modified and deleted by the exchange of the following messages:

– CREATE/MODIFY/DELETE BEARER REQUEST;

– CREATE/MODIFY/DELETE BEARER RESPONSE.

The DOWNLINK DATA NOTIFICATION message is sent by the SGW entity to the MME entity over the S11 interface.

The procedure follows the reception by the SGW entity of data from the PGW entity, with the mobile in ECM-IDLE mode. Upon receiving this message, the MME entity sends the S1-AP PAGING message to the eNB entities belonging to the TAC.

The MME entity may respond with the DOWNLINK DATA NOTIFICATION ACKNOWLEDGE message, indicating whether or not the request is accepted, or with the DOWNLINK DATA NOTIFICATION FAILURE INDICATION message if the mobile does not respond to the paging or if the mobile service request is rejected.

The CREATE INDIRECT DATA FORWARDING TUNNEL REQUEST/RESPONSE messages create a specific traffic bearer when handover occurs. This bearer forwards the data traffic received by the source eNB entity to the SGW entity to then be retransmitted to the mobile via the target eNB entity.

2.5.2. *Mobility management*

Mobility management messages are exchanged between the source and target MME entities, when the mobility of the mobile imposes a switch of MME entity.

The source MME entity sends the target MME entity the FORWARD RELOCATION REQUEST message, containing the context of the mobile.

The target MME entity responds with the FORWARD RELOCATION RESPONSE message when the resources needed for the handover have been reserved.

The response contains the values of the TEID parameter, which will enable the source SGW entity to redirect the traffic to the target SGW entity during handover.

Upon receiving this message, the source MME entity sends the source eNB entity the command to initiate handover.

The source MME entity sends the target MME entity the FORWARD ACCESS CONTEXT NOTIFICATION message to provide it with the elements of the context of the E-RAB bearer, such as the PDCP sequence number.

The target MME entity sends the source MME entity the FORWARD RELOCATION NOTIFICATION message to indicate that the handover procedure is complete.

The new MME entity sends the CONTEXT REQUEST message to the former one in the procedure of TAI updating, to retrieve information about the context of the mobile.

The former MME entity provides this information in the CONTEXT RESPONSE message, which may contain a positive or negative response. The new MME entity acknowledges this previous message with the CONTEXT ACKNOWLEDGE message.

2.6. DIAMETER protocol

2.6.1. *S6a interface*

Table 2.7 summarizes the DIAMETER messages exchanged over S6a interface between the MME and HSS entities.

AIR and AIA messages allow an MME entity to retrieve the authentication data of the mobile (RAND, RES, AUTN, K_{ASME}).

ULR and ULA messages allow:

– to provide the HSS entity with the identity of the MME entity who attached the mobile and information on the mobile;

– the MME entity to retrieve the profile data of the mobile.

PUR and PUA messages are used to inform the HSS entity that MME entity deleted the mobile profile, after a long inactivity.

CLR and CLA messages allow the HSS entity to delete the mobile profile stored in the MME entity.

IDR and IDA messages allow the HSS entity updating mobile profile stored in the MME entity.

DDR and DDA messages allow the HSS entity to remove elements of the mobile profile stored in the MME entity.

RSR and RSA messages allow to inform the MME entity of a restart of the HSS entity.

NOR and NOA messages allow to inform the HSS entity on certain events on the mobile.

Request	Response
Authentication-Information-Request (AIR)	Authentication-Information-Answer (AIA)
Update-Location-Request (ULR)	Update-Location-Answer (ULA)
Purge-UE-Request (PUR)	Purge-UE-Answer (PUA)
Insert-Subscriber-Data-Request (IDR)	Insert-Subscriber-Data-Answer (IDA)
Cancel-Location-Request (CLR)	Cancel-Location-Answer (CLA)
Delete-Subscriber-Data-Request (DDR)	Delete-Subscriber-Data-Answer (DDA)
Reset-Request (RSR)	Reset-Answer (RSA)
Notify-Request (NOR)	Notify-Answer (NOA)

Table 2.7. *DIAMETER messages over S6a interface*

2.6.2. *Gx interface*

Table 2.8 summarizes DIAMETER messages exchanged over Gx interface between Policy Charging and Rules Function (PCRF) and Policy and Charging Enforcement Function (PCEF) hosted in the PGW entity.

Request	Response
Credit-Control-Request (CCR)	Credit-Control-Answer (CCA)
Re-Auth-Request (RAR)	Re-Auth-Answer (RAA)

Table 2.8. *DIAMETER messages over Gx interface*

CCR and CCA messages enable the PCEF entity to solicit the PCRF in order to:

– retrieve the rules to apply to the default bearer created by the EPS network;

– inform the PCRF entity to the termination of the session on the EPS network.

RAR and RAA messages allow the PCRF entity to solicit the PCEF entity in order to provide the rules to be applied for the dedicated bearer.

2.6.3. *Rx interface*

Table 2.9 summarizes DIAMETER messages exchanged over Rx interface between the PCRF and AF (Application Function) entities.

Request	Response
Authenticate and Authorize Request (AAR)	Authenticate and Authorize Answer (AAA)
Re-Auth-Request (RAR)	Re-Auth-Answer (RAA)
Session Termination Request (STR)	Session Termination Answer (STA)
Abort-Session-Request (ASR)	Abort-Session-Answer (ASA)

Table 2.9. *DIAMETER messages over Rx interface*

AAR and AAA messages allow the AF entity to provide the characteristics of the media so that the EPS network can establish the dedicated bearer.

RAR and RAA messages allow the PCRF entity to notify the AF entity that a number of IP flows were disabled following a deactivation applied at the PCEF entity.

STR and STA messages allow the AF entity to inform the session is finished, so that the EPS network release the dedicated bearer.

ASR and ASA messages allow the PCRF entity to notify the AF entity that session on the EPS network is terminated, for example as a result of the loss of coverage of the mobile.

2.6.4. *S9 interface*

DIAMETER messages exchanged over R9 interface between Home PCRF (H-PCRF) and Visited PCRF (V-PCRF) are identical to the messages exchanged over Gx and Rx interfaces.

3

Mobile Network Procedures

3.1. Attachment

At the end of the connection procedure, the mobile starts the attachment procedure to the Evolved Packet System (EPS), which comprises the following steps:

– the mutual authentication between the User Equipment (UE) and the Mobility Management Entity (MME) corresponding to the AKA (Authentication and Key Agreement) mechanism;

– the security of Non-Access Stratum (NAS) messages;

– the location of the mobile related to the Tracking Area Identity (TAI) and to the E-UTRAN Cell Global Identifier (ECGI);

– the security of the radio interface LTE-Uu;

– the establishment of a default bearer for the Internet services. In the case where the voice or video conversational call is supported by the EPS network, a default bearer (QCI = 5) is established for transporting the signaling exchanged between the mobile and the IP Multimedia Subsystem (IMS);

– the granting of a Globally Unique Temporary Identity (GUTI).

Mutual authentication and the security for NAS message radio interface LTE-Uu are discussed in Chapter 8.

The establishment of the default bearer is described in section 3.3.1.

3.2. Location update

The location update is initiated by the mobile when entering a new location area or when its timer expires.

Before starting the procedure, the mobile is in the following state:

– EMM-REGISTERED: the mobile is attached to the MME entity;

– RRC-IDLE: the mobile is not connected to the evolved Node Base station (eNB);

– ECM-IDLE: the connection between the mobile and the MME entity is released;

During the procedure, the mobile moves to the following state:

– EMM-REGISTERED: the mobile remains attached to the MME entity;

– RRC-CONNECTED: the mobile is connected to the eNB entity;

– ECM-CONNECTED: the connection between the mobile and MME entity is established.

After the procedure, the mobile returns to the initial state.

The update location results in the following (Figure 3.1):

– the transfer of the mobile context of the former to the new MME entity, if the new area belongs to a new group (pool);

– the mutual authentication between the mobile and new MME entity;

– the establishment of the default bearer between the new Serving Gateway (SGW) and the PDN Gateway (PGW), if the mobile changes group;

– the update of the HSS entity with the identity of the new MME entity.

1) The mobile initializes the update location by sending the EMM TRACKING AREA UPDATE REQUEST message containing the temporary identity GUTI and the last visited area (TAI).

The EMM TRACKING AREA UPDATE REQUEST message is carried:

– by the RRC *ConnectionSetupComplete* message on the radio interface LTE-Uu between the mobile and the eNB entity. The RRC message contains the Globally Unique MME Identity (GUMMEI) of the entity who attached the mobile. This identity is extracted from the temporary identity (GUTI).

– by the S1-AP INITIAL MESSAGE UE message on the S1-MME interface between the eNB and MME entities. In the case of a group change, the new MME entity is determined by the eNB entity.

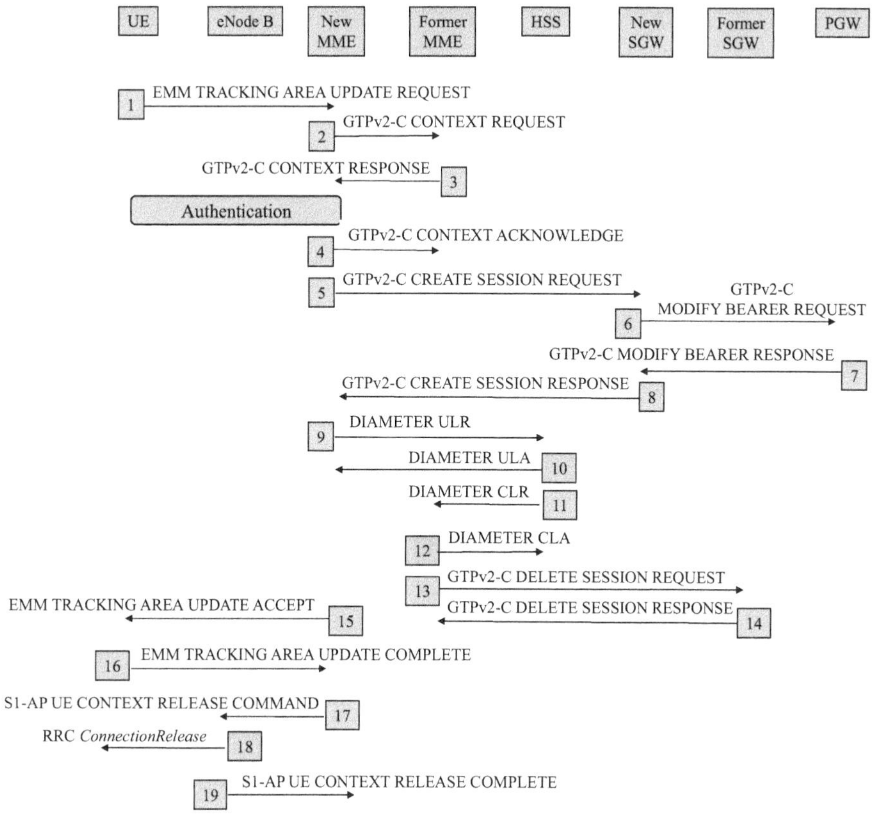

Figure 3.1. *Location update procedure*

2) 3) The new MME entity retrieves the identity of the former MME entity from the temporary identity (GUTI) of the mobile and sends the

GTPv2-C CONTEXT REQUEST message. The former MME entity returns the context of the mobile in the GTPv2-C CONTEXT RESPONSE message.

4) The authentication procedure is performed between the mobile and the new MME entity, which then acknowledges the transfer of the context by sending the GTPv2-C CONTEXT ACKNOWLEDGE message to the former MME entity.

5) 8) The new MME entity creates a session with a new SGW entity through an exchange of GTPv2-C messages: CREATE SESSION REQUEST/RESPONSE. The request contains the IP address of the PGW entity recovered from the context of mobile and mobile location identifiers (TAI and ECGI).

6) 7) The new SGW entity changes the S5 bearer with the PGW entity through an exchange of GTPv2-C messages: MODIFY BEARER REQUEST/RESPONSE.

9) 10) The new MME entity updates the location of the mobile to the HSS entity through the exchange of DIAMETER messages: ULR (Update-Location Request) and ULA (Update-Location Answer).

11) 12) For its part, the HSS entity shall cancel the mobile information in the former MME through the exchange of DIAMETER messages: CLR (Cancel-Location-Request) and CLA (Cancel-Location-Answer).

13) 14) On receipt of this message, the former MME removes the context of the mobile and does the same with the SGW entity by removing the context by the exchange of GTPv2-C messages: DELETE SESSION REQUEST/RESPONSE.

15) The new MME entity can then respond to the mobile by the transmission of the EMM TRACKING AREA UPDATE ACCEPT message containing a new temporary identity (GUTI).

The EMM TRACKING AREA UPDATE ACCEPT message is carried:

– by the S1-AP DOWNLINK NAS TRANSPORT message on the S1-MME interface between the eNB and MME entities;

– by the RRC *DLInformationTransfer* message on the radio interface LTE-Uu between the mobile and the eNB entity.

16) The EMM TRACKING AREA UPDATE ACCEPT message is acknowledged by the mobile through the EMM TRACKING AREA UPDATE COMPLETE message carried:

– by the RRC *ULInformationTransfer* message on the radio interface LTE-Uu, between the mobile and the eNB entity;

– by the S1-AP UPLINK NAS TRANSPORT message on the S1-MME interface between the eNB and the MME entities.

17) The new MME controls the release of the resources from the eNB entity by sending the S1-AP EU CONTEXT RELEASE COMMAND message.

18) The eNB entity controls the release of the mobile resources by sending the RRC *ConnectionRelease* message, which causes the mobile to move to the RRC-IDLE/ECM-IDLE states.

19) The eNB entity responds to the new MME entity with the S1-AP EU CONTEXT RELEASE COMPLETE message, which causes the MME entity to move to the ECM-IDLE state.

3.3. Session establishment

The EPS network carries traffic data (IP packets) transparently to the PGW entity that performs IP packet routing.

The IP packets are carried in bearers constructed between EPS network entities (Figure 3.2).

The Data Radio Bearer (DRB) is constructed between the User Equipment (UE) and the eNB entity. The RRC (Radio Resource Control) signaling, which is exchanged between the mobile and the eNB entity, is responsible for constructing this bearer.

The S1 bearer is constructed between the eNB and SGW entities. The S1-AP signaling, exchanged between the eNB and MME entities, and the GTPv2-C (GPRS Tunneling Protocol Control plane) signaling, exchanged between the MME and SGW entities, are responsible for constructing this bearer.

The S5 bearer is constructed between the SGW and PGW entities. The GTPv2-C signaling exchanged between the SGW and PGW entities is responsible for constructing this bearer.

The connection of the radio bearer and the S1 bearer, performed by the eNB entity, constitutes the EPS Radio Access Bearer (E-RAB).

The connection of the E-RAB and S5 bearers, which is performed by the SGW entity, constitutes the EPS bearer.

The transport of IP packets (flows) in different bearers (DRB, S1, S5) is described in Figure 3.3.

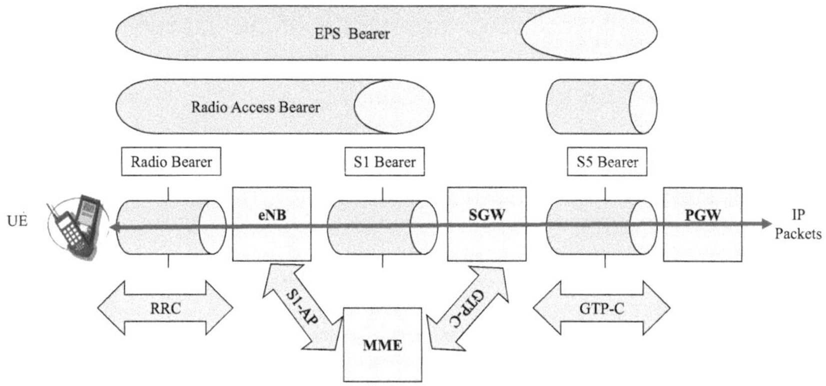

Figure 3.2. *Construction of the bearers*

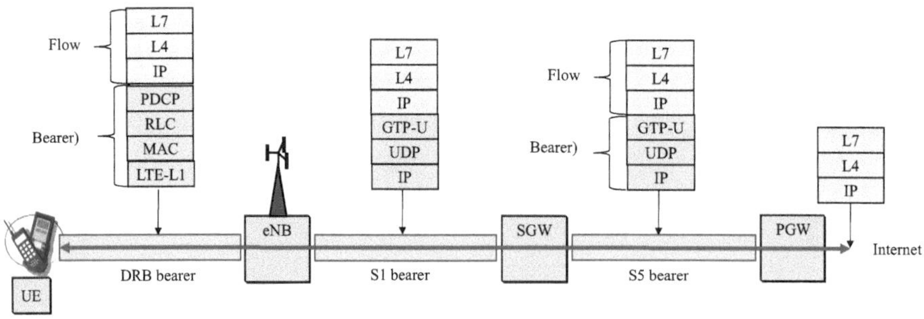

Figure 3.3. *Transport of IP packets in bearers*

The IP transport network that enables communication between the EPS network entities routes the S1 or S5 bearers.

The PGW entity is the only entity of the mobile network that routes traffic data (flows).

The eNB and SGW entities do not perform routing. They only provide the connection between the bearers.

There are two types of bearers in the EPS network:

– the default bearer, established when attaching the mobile, used, for example, to transport SIP signaling or Internet services;

– the dedicated bearer, established following a specific request from the mobile, used, for example, for transport of RTP (Real-time Transport Protocol) streams containing voice or video.

3.3.1. *Establishment of the default bearer*

The establishment of the default bearer is described in Figure 3.4.

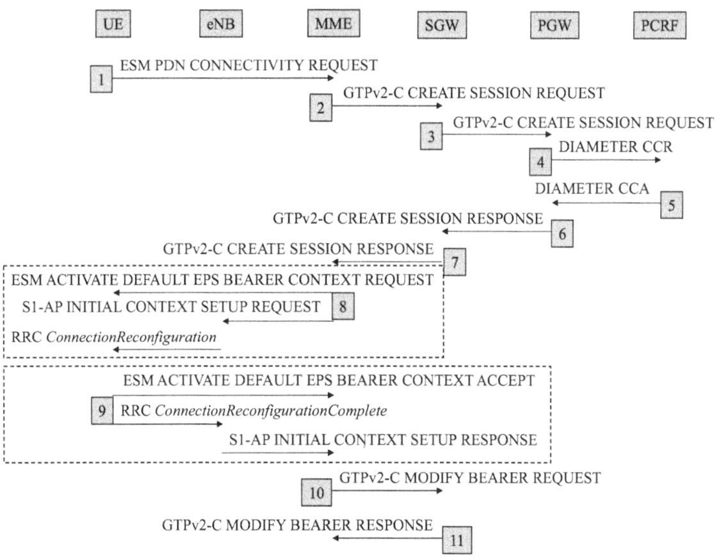

Figure 3.4. *Establishment of the default bearer*

1) The establishment of the default bearer is triggered by the mobile when it sends the ESM PDN CONNECTIVITY REQUEST message to the MME entity.

The MME entity selects the Serving Gateway (SGW) in its group and the PDN Gateway (PGW) from a DNS (Domain Name Server) resolution of the APN.

2) The MME entity sends the GTPv2-C CREATE SESSION REQUEST message to create a context at the SGW entity.

The GTPv2-C CREATE SESSION REQUEST message contains the IP (Internet Protocol) address of the PGW entity, the APN and the default bearer profile.

3) The SGW entity sends the GTPv2-C CREATE SESSION REQUEST message to create a context at the PGW entity.

The GTPv2-C CREATE SESSION REQUEST message contains the Tunnel Endpoint Identifier (TEID) that the PGW entity will use at the GPRS Tunneling Protocol User plane (GTP-U) level for the S5 bearer.

4) The PGW entity sends to the Policy and Charging Rules Function (PCRF) the DIAMETER CCR (Credit-Control-Request) message for authorization to open the default bearer.

The PCRF entity compares the profile of the mobile with the rules defined for the network and stored in the Subscription Profile Repository (SPR) database.

5) 13) The PCRF entity responds to the PGW entity with a DIAMETER CCA (Credit-Control-Answer) message containing the rules to be applied to the default bearer (filter parameters, charging mode).

6) The PGW entity responds to the SGW entity with the GTPv2-C CREATE SESSION RESPONSE message, which contains the following information:

– the TEID identifier that the SGW entity will use at the GTP-U protocol level for the S5 bearer;

– the mobile configuration (IP address of the mobile, IP address of its DNS server).

7) The SGW entity responds to the MME entity with a GTPv2-C CREATE SESSION RESPONSE, which contains the following information:

– the TEID identifier that the eNB entity will use at the GTP-U protocol level for the S1 bearer;

– the mobile configuration (IP address of the mobile, IP address of its DNS server).

8) The MME responds to the mobile with the ESM ACTIVATE DEFAULT EPS BEARER CONTEXT REQUEST message supported by:

– the S1-AP INITIAL CONTEXT SETUP REQUEST message that allows the creation of the mobile context at the eNB entity and that contains the identifier TEID assigned by the SGW entity;

– the RRC *ConnectionReconfiguration* message that indicates the LCID (Logical Channel Identifier) value of the DRB bearer.

The ESM ACTIVATE DEFAULT EPS BEARER CONTEXT REQUEST message contains the mobile configuration (IP address of mobile, IP address of its DNS server).

9) The mobile acknowledges the activation of the default bearer with the ESM ACTIVATE DEFAULT EPS BEARER CONTEXT ACCEPT message, which is carried by the RRC *ConnectionReconfigurationComplete* message on the radio interface LTE-Uu and the S1-AP INITIAL CONTEXT SETUP RESPONSE message on the S1-MME interface.

The S1-AP INITIAL CONTEXT SETUP RESPONSE message contains the TEID identifier which the SGW entity will use at the GTP-U protocol level for the S1 bearer.

The RRC *ConnectionReconfigurationComplete* message acknowledges the reception of the RRC *ConnectionReconfiguration* message.

10) The MME entity transfers the TEID identifier received from the eNB entity to the SGW entity in the GTPv2-C MODIFY BEARER REQUEST message.

11) The SGW entity acknowledges the message received by the GTPv2-C MODIFY BEARER RESPONSE message.

3.3.2. *Re-establishment of the default bearer*

The re-establishment of the default bearer is initiated from a service request. The procedure is triggered by the mobile in idle state when the first IP packet is outgoing or by the SGW entity when the first IP packet is incoming in order to activate the DRB and S1 bearers.

3.3.2.1. *Re-establishment initiated by the mobile*

The re-establishment of the default bearer initiated by the mobile is described in Figure 3.5.

1) The mobile starts the service request by activating the RRC connection and then by sending a EMM SERVICE REQUEST message to the MME entity.

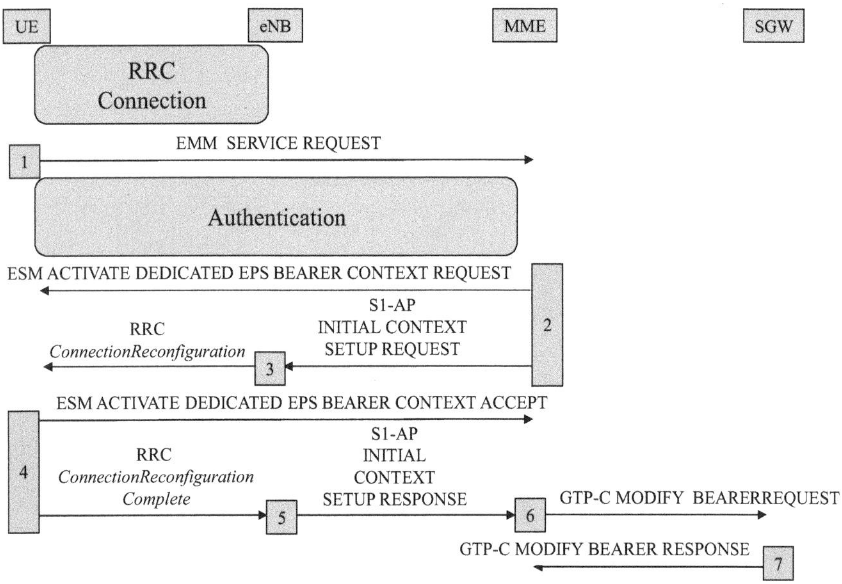

Figure 3.5. *Re-establishment initiated by the mobile*

2) 5) After the authentication phase, the MME entity triggers the activation of the bearer on the S1-U interface by exchanging with the eNB entity the S1-AP messages: INITIAL CONTEXT SETUP REQUEST/RESPONSE.

In the S1-AP INITIAL CONTEXT SETUP REQUEST message, the MME indicates the value of the TEID identifier that the eNB entity will use for the S1 bearer. This value has been saved in the context of the MME entity.

In the S1-AP INITIAL CONTEXT SETUP RESPONSE message, the eNB entity indicates the value of the TEID identifier that the SGW entity will use for the S1 bearer.

3) 4) The eNB entity establishes the bearer on the radio interface LTE-Uu by exchanging RRC messages: RRC *ConnexionReconfiguration* and RRC *ConnexionReconfigurationComplete*.

In the RRC *ConnexionReconfiguration* message, the eNB entity indicates the LCID value of the DRB bearer.

The ESM ACTIVATE DEFAULT EPS BEARER CONTEXT REQUEST message is carried by the S1-AP INITIAL CONTEXT SETUP REQUEST message on the S1-MME interface and by the RRC *ConnectionReconfiguration* message on the radio interface LTE-Uu.

The ESM ACTIVATE DEFAULT EPS BEARER CONTEXT ACCEPT message is carried by RRC *ConnectionReconfigurationComplete* on the radio interface LTE-Uu and by the S1-AP INITIAL CONTEXT SETUP RESPONSE message on the S1-MME interface.

6) 7) The MME entity triggers the activation of the bearer on the S1-U interface by exchanging with the SGW entity the GTPv2-C messages: MODIFY BEARER REQUEST/RESPONSE.

In the GTPv2-C MODIFY BEARER REQUEST message, the MME forwards the TEID identifier that the SGW entity will use for the S1 bearer.

3.3.2.2. *Re-establishment initiated by the SGW entity*

The re-establishment of the default bearer initiated by the SGW entity is described in Figure 3.6.

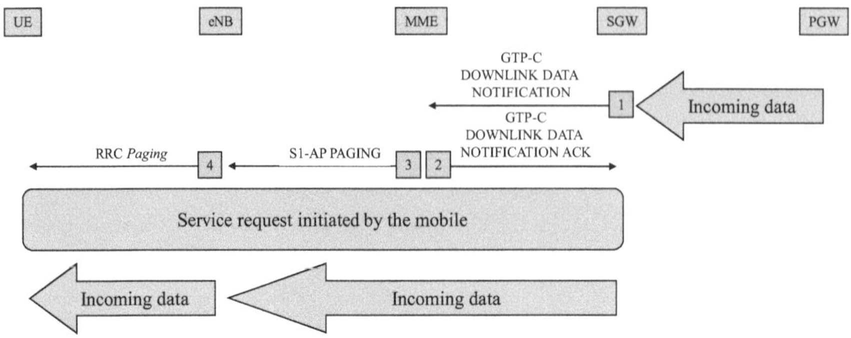

Figure 3.6. *Re-establishment initiated by the SGW entity*

As the S5 bearer is active, the SGW entity receives data from the PGW entity while the mobile is in the idle state and the DRB and S1 bearer are not established:

1) 2) The SGW entity generates the GTPv2-C DOWNLINK DATA NOTIFICATION message to the MME entity, acknowledged by the GTPv2-C NOTIFICATION DATA DOWNLINK ACK message.

3) The MME entity starts the paging procedure by forwarding to the eNB entities of the location area (TAI), the S1-AP PAGING message containing a subset (S-TMSI) of the temporary identity (GUTI) of the mobile.

4) Each eNB entity broadcasts the RRC *Paging* message on the radio interface LTE-Uu.

On receipt of this message, the mobile initiates the procedure of service request.

When the DRB and S1 bearers are activated, the SGW entity transfers the incoming data stored to the eNB entity, which in turn transmits them to the mobile.

3.3.3. *Establishment of the dedicated bearer*

The establishment of the dedicated bearer is triggered by the AF entity (IMS network in the case of telephone or videophone services). The link between the AF entity and the EPS network is provided by the PCRF entity that transfers to the PGW entity the characteristics of the dedicated bearer (voice or video) that it is necessary to establish.

The dedicated support is coupled to the default bearer assigned to the telephone signaling, in that the terminations are the same for both types of bearer.

The establishment of the dedicated bearer is described in Figure 3.7.

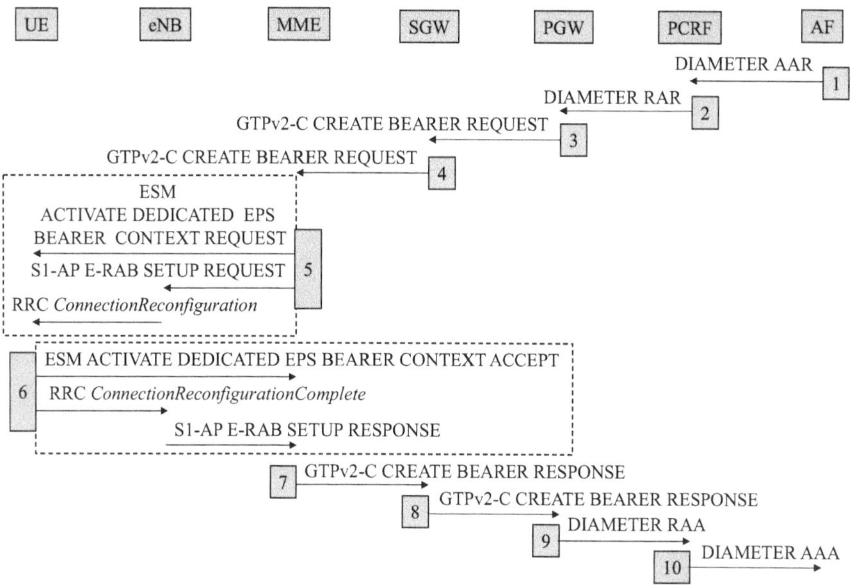

Figure 3.7. *Establishment of the dedicated bearer*

1) The establishment of the dedicated support is triggered by the AF entity when sending to the PCRF entity the DIAMETER AAR (Authenticate-Authorize-Request) message.

2) The PCRF entity consults its SPR database to retrieve the traffic profile to apply and transmits to the PGW entity the DIAMETER RAR (Re-Auth-Request) message containing the flow characteristics to establish.

3) The PGW entity sends to the SGW entity the GTPv2-C CREATE BEARER REQUEST message containing the TEID identifier of the S5

bearer that the SGW entity will need to use in the GTP-U header when sending traffic to the PGW entity.

4) The SGW entity sends to the MME entity the GTPv2-C CREATE BEARER REQUEST message containing the TEID identifier of the S1 bearer that the eNB entity will need to use in the GTP-U header when sending traffic to the SGW entity.

5) The MME entity sends to the mobile the ESM ACTIVATE DEDICATED EPS BEARER CONTEXT REQUEST message carried by the S1-AP E-RAB SETUP REQUEST message on the S1-MME interface and by the RRC *ConnectionReconfiguration* message on the LTE-Uu interface.

The S1-AP E-RAB SETUP REQUEST message contains the TEID identifier that the MME entity has received from the SGW entity.

The RRC *ConnectionReconfiguration* message contains the Logical Channel Identifier (LCID) of the dedicated bearer.

6) The mobile responds to the MME entity with the ESM ACTIVATE DEDICATED EPS BEARER CONTEXT ACCEPT message carried by the RRC *ConnectionReconfigurationComplete* message on the LTE-Uu interface and by the S1-AP E-RAB SETUP RESPONSE message on the S1-MME interface.

The S1-AP E-RAB SETUP RESPONSE message contains the TEID identifier of the S1 bearer that the SGW entity will need to use in the GTP-U header when it sends traffic to the eNB entity.

7) The MME entity responds to the SGW entity with the GTPv2-C CREATE BEARER RESPONSE message containing the TEID identifier that the MME entity has received from the eNB entity.

8) The SGW entity responds to the PGW entity with the GTPv2-C CREATE BEARER RESPONSE message containing the TEID identifier of the S5 bearer that PGW entity will need to use in the GTP-U header when it sends traffic to the SGW entity.

9) The PGW entity responds to the PCRF entity with the DIAMETER RAA (Re-Auth Answer) message.

10) The PCRF entity responds to the AF entity with the DIAMETER AAA (Authenticate-Authorize-Answer) message.

3.4. Handover

The handover is the mechanism that maintains the current session when the mobile changes the radio cell.

During the intrasystem handover, cell change occurs without mobile network changes.

During the intrasystem handover, the SGW or PGW entities play the role of an anchor point, which will hide the terminal mobility to the destination of the IP packets.

There are two types of intrasystem handover:

– X2-based handover from the name of the interface between the eNB entities;

– S1-based handover from the name of the interface between the MME and eNB entities.

X2-based handover occurs if both eNB entities, the source and the target belong to the same group (pool) and the X2-AP interface has been enabled.

During X2-based handover, the MME entity does not change, which is not the case for an SGW entity that can be relocated or not.

S1-based handover occurs for the following cases:

– X2-AP interface between two eNB entities of the same group was not activated, the MME must not be relocated and the SGW entity may be relocated or not;

– both eNB entities belong to two different groups, the MME entity must be relocated and the SGW entity may be relocated or not.

During the intersystem handover, cell change is accompanied by a change of network, the mobile being transferred from 4G network to 2G or 3G networks.

There are two types of intersystem handover:

– PS-PS (Packet-Switched) handover, for which the mobile does not change the mode when changing the network;

– PS-CS (Circuit-Switched) handover, for which the mobile is also changing the mode.

The handover procedure is activated by the source eNB entity from measurements on the radio signal made by:

– the eNB entity for the uplink direction;

– the mobile for the downlink direction for the serving and neighboring cells. These measures are recovered by the eNB entity in RRC *MeasurementReport* messages.

The handover procedure takes place in three phases:

– the preparation phase corresponding to the decision of cell change and resource reservation;

– the execution phase corresponding to the mobile connection to the target eNB entity;

– the completion phase corresponding to the establishment of final bearers and the release of the old resources.

The handover is carried out with a disconnection of the mobile from the source eNB entity and a reconnection of the mobile to the target base station.

During the execution of the handover, the incoming data are forwarded to the target base station that stores and delivers it to the mobile when it connects.

3.4.1. *Handover based on X2*

The functional architecture of the handover based on X2 without relocation of the SGW entity is described in Figure 3.8.

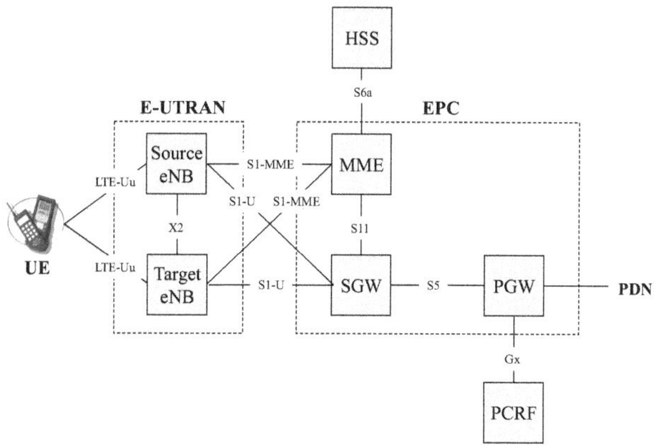

Figure 3.8. *Handover based on X2 without relocation Functional architecture*

The establishment of the bearers during the phases of preparation, execution and completion is described in Figure 3.9.

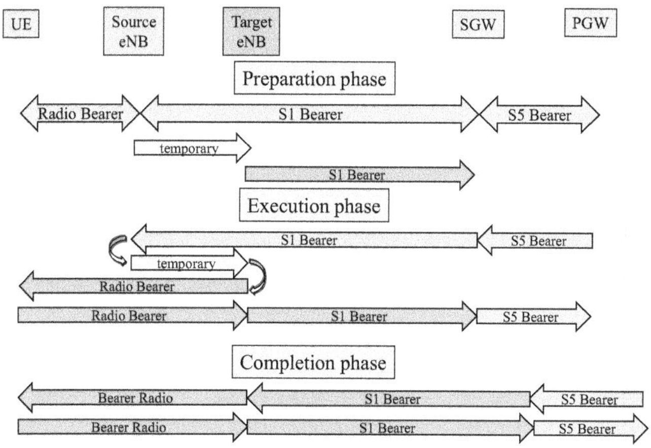

Figure 3.9. *Handover based on X2 Bearer establishment*

Following the preparation phase, the following bearers are established but not used:

– bearer from the source eNB entity to the target eNB entity, which is temporary and unidirectional;

– the new S1 bearer from the target eNB entity to the SGW entity.

At the end of the execution phase, the bearers used are:

– PGW–SGW (S5 bearer of origin), SGW–source eNB (S1 bearer of origin), source eNB–target eNB and target eNB–UE (DRB bearer) for incoming data;

– UE–target eNB (DRB bearer), target eNB–SGW (new S1 bearer) and SGW–PGW (S5 bearer of origin) for the outgoing data.

Following the completion phase, the bearers used for the incoming data are: PGW–SGW (S5 bearer of origin), SGW–target eNB (new S1 support) and target eNB–UE (DRB bearer).

3.4.2. *Handover based on S1*

The functional architecture of the handover based on S1 with relocation of the MME and SGW entities is described in Figure 3.10.

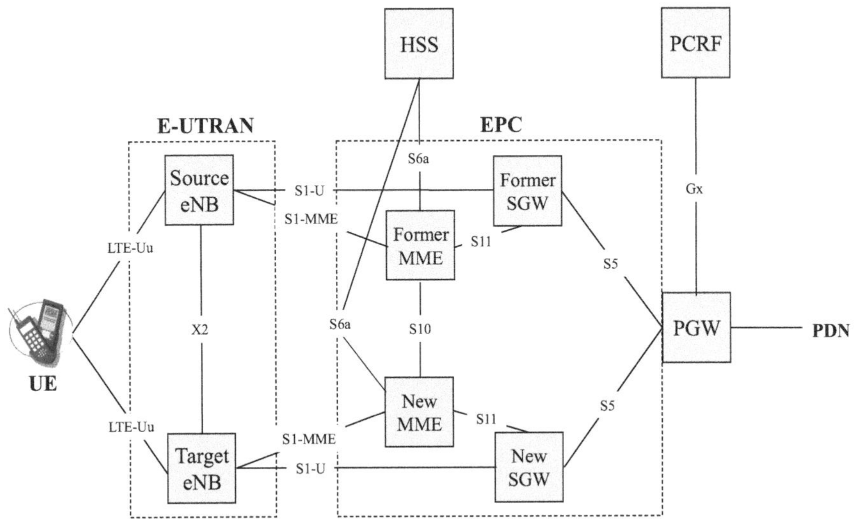

Figure 3.10. *Handover based on S1 with relocation Functional architecture*

The establishment of the bearers during the phases of preparation, execution and completion is described in Figure 3.11.

Following the preparation phase, the following bearers are established but not used:

– the new S5 bearer from the new SGW entity to the PGW entity;

– the new S1 bearer from the target eNB entity to the new SGW entity;

– the unidirectional and temporary bearer from the source eNB entity to the former SGW entity;

– the unidirectional and temporary bearer from the former SGW entity to the new SGW entity;

– the unidirectional and temporary bearer from the new SGW entity to the target eNB entity.

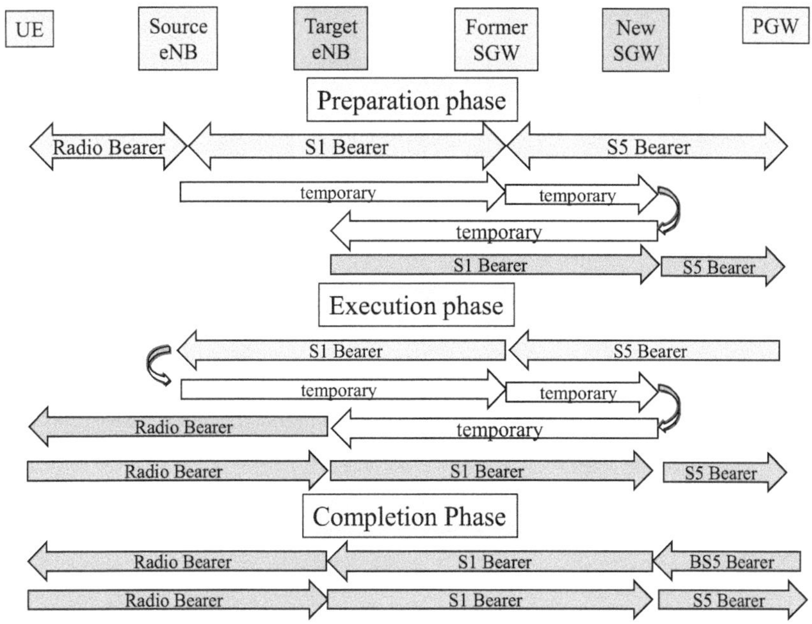

Figure 3.11. *Handover based on S1 Bearer establishment*

At the end of the execution phase, the bearers used are:

– PGW–former SGW (S5 bearer of origin), former SGW–source eNB (S1 bearer of origin), source eNB–former SGW (temporary bearer), former

SGW–new SGW (temporary bearer), new SGW–target eNB (temporary bearer) and target eNB–UE (DRB bearer) for the incoming data;

– UE–target eNB (DRB bearer), target eNB–new SGW (new S1 bearer) and new SGW–PGW (new S5 bearer) for the outgoing data.

Following the completion phase, the bearers used for the incoming data are: PGW–new SGW (new S5 bearer), new SGW–target eNB (new S1 support) and target eNB–UE (DRB bearer).

4

Transport Network MPLS-VPN Technology

4.1. MPLS network

4.1.1. *Architecture*

In a MPLS (Multi-Protocol Label Switching) network, IP (Internet Protocol) packets are transferred from a label (or identifier) associated with a destination address that corresponds to a Forwarding Equivalence Class (FEC).

A label can be generated for all interfaces of a router or for each interface of a router.

The labels are used to build a virtual circuit, called Label Switching Path (LSP), and have only a local significance, not an end-to-end meaning.

The basic principle of the MPLS is the switching of labels, which are simple integers inserted between the Ethernet (layer 2) header and IP (layer 3) header.

The labels allow routers to make a switch from these labels without having to check the IP header and the routing table.

This technique is similar to the cell switching for ATM (Asynchronous Transfer Mode) technology, with the Virtual Path Identifier (VPI) and Virtual Channel Identifier (VCI) or virtual circuit switching for a Frame Relay technology with the Data Link Connection Identifier (DLCI).

The MPLS network is built from two types of routers (Figure 4.1):

– Edge Label Switching Router (LSR) located on the edge of the MPLS network. Edge LSR is called ingress Edge LSR for the incoming flow, and egress Edge LSR for the outgoing flow.

– The intermediate router between Edge LSR is called Core LSR.

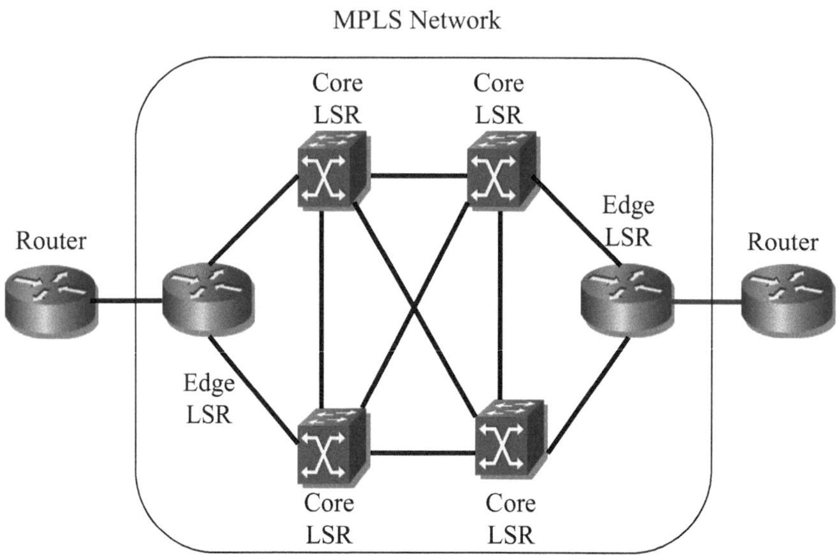

Figure 4.1. *MPLS network architecture*

Edge LSR performs IP routing (Figure 4.2). Ingress Edge LSR inserts the label (PUSH function), and egress Edge LSR removes the label (POP function).

Core LSR performs the label switching (SWAP function) (Figure 4.2).

4.1.2. *LSR tables*

The information required to reach the remote Edge LSR is stored in the Routing Information Base (RIB) located in the control plane of LSR. The RIB table is advised by the OSPF (Open Shortest Path First) or IS-IS (Intermediate System to Intermediate System) routing protocols (Figure 4.3).

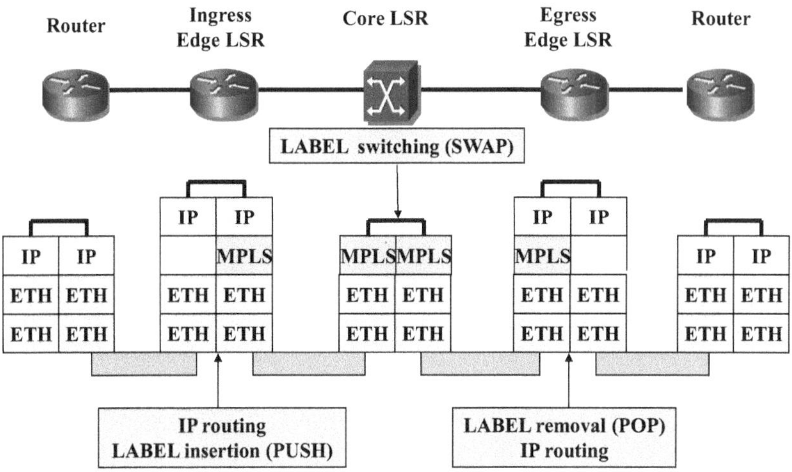

Figure 4.2. *Protocol architecture of the MPLS network*

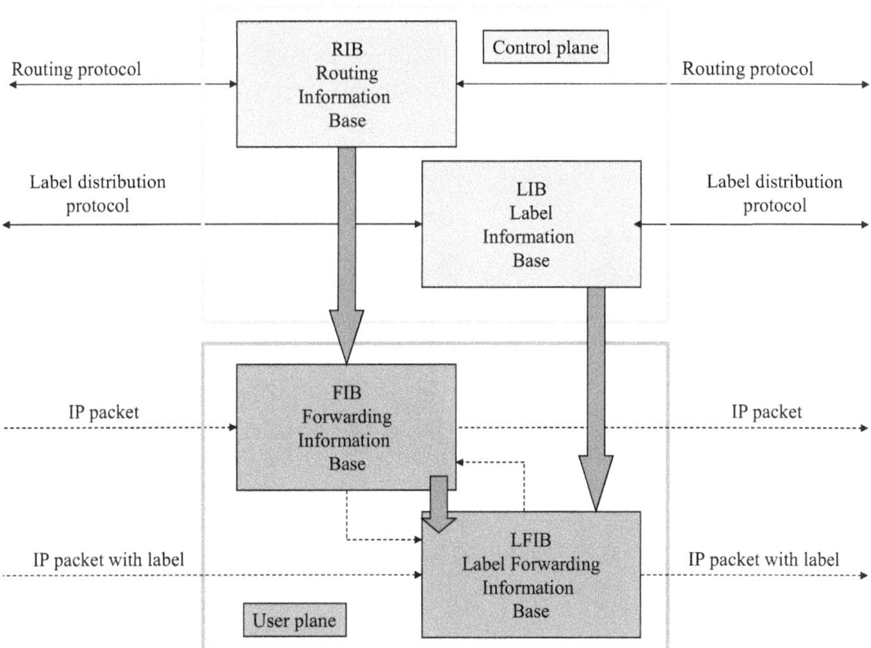

Figure 4.3. *LSR tables. For a color version of this figure, see www.iste.co.uk/perez/4gmobile.zip*

The information stored in the RIB table is transferred to another table, the Forwarding Information Base (FIB) located in the user plane of the LSR. The FIB table allows also to perform an aggregation of different tables used for IP and Ethernet mechanisms (Figure 4.3).

The Label Information Base (LIB) is filled by the Label Distribution Protocol (LDP). It contains the links between input and output labels for each IP address of LSRs. The LIB table is located in the control plane of the LSR (Figure 4.3).

The Label Forwarding Information Base (LFIB) is an aggregation of the FIB and LIB tables. It is used by the Core LSR for switching labels by the ingress Edge LSR to insert the label and by the egress LSR Edge router to remove the label. The LFIB table is located in the user plane of LSRs (Figure 4.3).

4.1.3. *PHP function*

The Penultimate Hop Popping (PHP) is to remove the MPLS header between the egress Edge LSR and the upstream Core LSR (Figure 4.4).

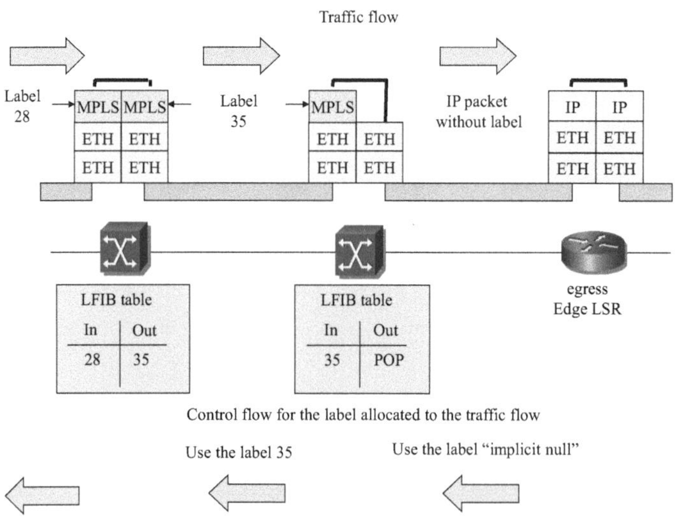

Figure 4.4. *PHP function*

If the packets received by the egress Edge LSR are labeled, it must first analyze the LFIB table to remove the label, then the FIB routing table to find the output interface.

The operation in the LFIB table is unnecessary. The egress LSR Edge announces the implicit-null label to its upstream routers. An LSR as having the value "implicit-null" for the output label removes the label.

4.1.4. *MPLS header*

The MPLS header has a size of 4 bytes and contains the following fields (Figure 4.5):

– Label: this field includes the value of the label assigned to the packet.

– EXP (Experimental) or TC (Traffic Class): this field includes the value of the traffic class to which the packet belongs.

– S (Stack): this field is used to stack the MPLS headers. The bit to ZERO indicates that the following header is an MPLS header. A bit to ONE indicates that the stack of MPLS headers is completed.

– TTL (Time To Live): this field has the same role as the TTL field in the IP header.

```
 0                   1                   2                   3
 0 1 2 3 4 5 6 7 8 9 0 1 2 3 4 5 6 7 8 9 0 1 2 3 4 5 6 7 8 9 0 1
+-+-+-+-+-+-+-+-+-+-+-+-+-+-+-+-+-+-+-+-+-+-+-+-+-+-+-+-+-+-+-+-+
|                LABEL                  | Exp |S|       TTL     |
+-+-+-+-+-+-+-+-+-+-+-+-+-+-+-+-+-+-+-+-+-+-+-+-+-+-+-+-+-+-+-+-+
```

Figure 4.5. *Format of the MPLS header*

When the ingress Edge LSR inserts the label, the TTL field of the MPLS header can be set to the value of the TTL field in the IP header, which must be previously decremented by a unit.

When a label is swapped by the Core LSR, the value of the TTL field of the MPLS header is then decremented in turn by one.

When a label is removed by the egress Edge LSR, the value of the TTL field in the IP header, before routing, is replaced with the value of the TTL field of the MPLS header.

However, there may be situations where it is preferred to a decrease in one unit the value of the TTL field in the IP header only at the Edge routers LSR.

4.1.5. *LDP protocol*

4.1.5.1. *Principles of the labeling*

LDP allows the exchange of messages between adjacent LSRs for assigning a label to the IP address of each LSR. There are four types of LDP messages:

– discovery message detects the adjacent LSR;

– initialization message establishes the LDP session between two LSRs;

– announcement message provides the correspondence between the label and the IP address of the LSR;

– notification message reports the error.

At first, LSR transmits discovery messages. These messages are encapsulated in a UDP header. The UDP segment is encapsulated in an IP header, for which the destination address is the multicast address 224.0.0.2. Adjacent LSR responds also with a discovery message.

During the exchanges of messages of discovery, each LSR learns the identifier of the LSRs. LSR with the highest identifier is the active LSR for each pair. The active LSR performs the opening of the TCP connection, the destination port number being equal to 646.

The active LSR then sends the initialization message containing the parameters of the LDP. It waits 15 seconds for the response of the adjacent router. If negative, it returns the message with new parameters and waits 30 seconds for the response of the adjacent router. For each negative response, the active router doubles timer until the value of 2 minutes. This value is retained for the following exchanges.

The method of label distribution called downstream indicates that the LDP messages containing the label associated with the IP address of the

LSR are propagated from the closest router for the IP address to reach to the router farther (downstream to upstream).

The downstream method has two variants: Unsolicited Downstream and Downstream On Demand (Figure 4.6). In the first variant, the downstream LSRs systematically propagate all their labels to their neighbors upstream. In the second variant, the upstream LSRs explicitly ask downstream LSRs to provide a label for the requested IP address.

The establishment of the LSP can be controlled in independent or ordered mode (Figure 4.6). In the independent control mode, each LSR freely decides to link a label to the IP address and distributes the correspondence to its neighbors. In the ordered control mode, an LSR only binds a label to an IP address if it is the egress Edge LSR or if it has already received from next hop LSR (downstream router), a correspondence between the label and the IP address.

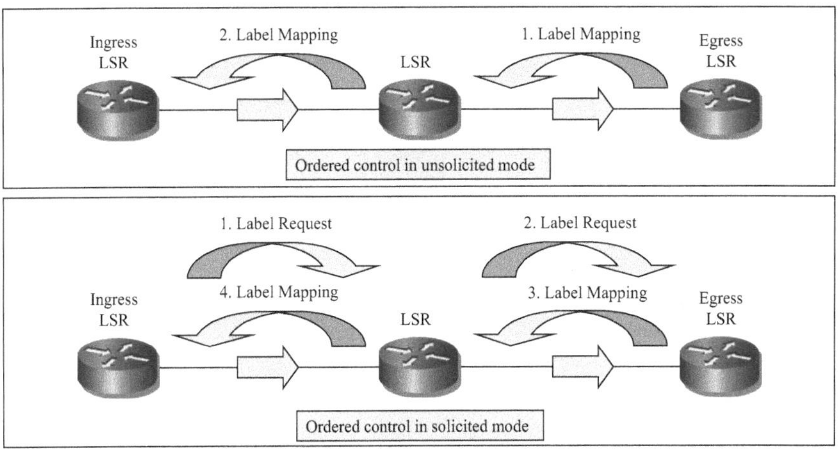

Figure 4.6. *LSP establishment modes*

If an LSR supports liberal label retention, it keeps all correspondence between the label and the IP address it receives from adjacent LSRs, whether or not downstream routers (Figure 4.7).

If an LSR supports conservative label retention, it retains only the association between the label and the IP address from the next hop LSR (downstream router) (Figure 4.7).

With the liberal label retention, LSR can transfer data as soon as the routing protocol has converged following a modification of the MPLS network. By cons, it must maintain a table with all received announcements.

LDP uses two mechanisms to detect loops. The first uses a counter incremented by one when the LDP message is processed by an LSR. When the maximum value is reached, the message is deleted. The second is to record the addresses of LSRs traversed. If an LSR is recognized in the list, the message is deleted.

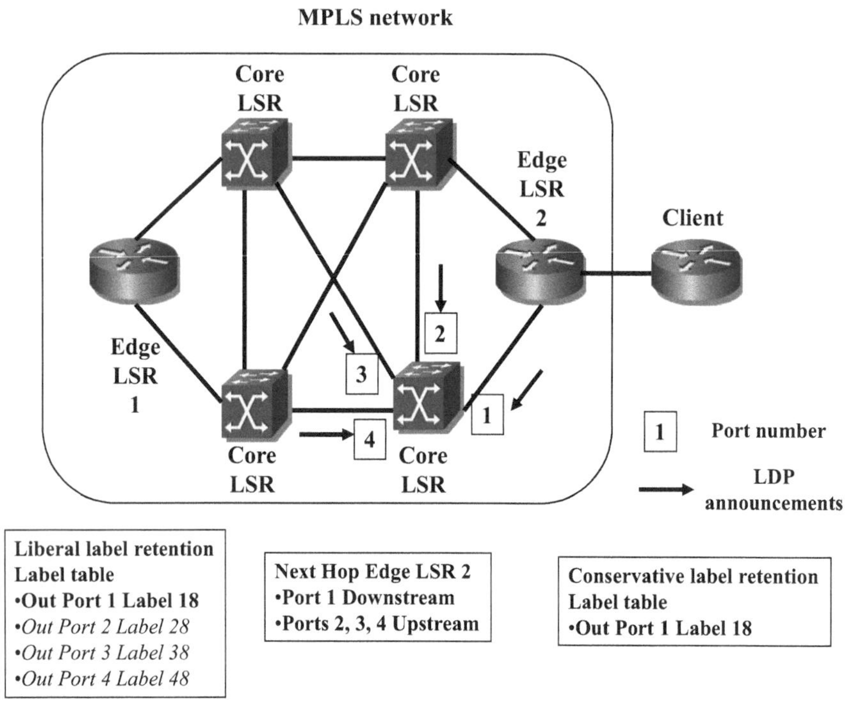

Figure 4.7. *Label retention*

4.1.5.2. *LDP PDU format*

The Protocol Data Unit (PDU) comprises a header and one or more LDP messages. The header contains the following fields (Figure 4.8):

– Version: this field inserts the LDP protocol version number (value of 1).

– PDU Length: this field inserts the length of the protocol data unit.

– LDP Identifier: this field inserts the identifier of LSR (IP address) and the label space used by the LSR.

```
 0                   1                   2                   3
 0 1 2 3 4 5 6 7 8 9 0 1 2 3 4 5 6 7 8 9 0 1 2 3 4 5 6 7 8 9 0 1
+-+-+-+-+-+-+-+-+-+-+-+-+-+-+-+-+-+-+-+-+-+-+-+-+-+-+-+-+-+-+-+-+
|          Version              |         PDU Length            |
+-+-+-+-+-+-+-+-+-+-+-+-+-+-+-+-+-+-+-+-+-+-+-+-+-+-+-+-+-+-+-+-+
|                 LDP Identifier (router ID)                    |
+-+-+-+-+-+-+-+-+-+-+-+-+-+-+-+-+-+-+-+-+-+-+-+-+-+-+-+-+-+-+-+-+
| LDP Identifier (label space) |
+-+-+-+-+-+-+-+-+-+-+-+-+-+-+-+-+
```

Figure 4.8. *Header of the PDU LDP format*

Each type of LDP message contains the following fields (Figure 4.9):

– U (Unknown): on receipt of an unknown message:

- if the bit is set to ZERO, the receiver must send a notification message;

- if the bit is set to ONE, the message is ignored.

– Message Type: this field inserts the LDP message type (Table 4.1).

– Message Length: this field inserts the size of the LDP message.

Message type	Description
HELLO	Discovery of the adjacent LSR
INITIALIZATION	LDP session establishment (parameter definition)
KEEPALIVE	Maintenance of the session in the absence of message exchange
ADDRESS	IP address of the next hop
ADDRESS_WITHDRAW	IP address withdraw of the next hop
LABEL_MAPPING	Mapping between the label and the FEC

Table 4.1. *LDP message types*

```
 0                   1                   2                   3
 0 1 2 3 4 5 6 7 8 9 0 1 2 3 4 5 6 7 8 9 0 1 2 3 4 5 6 7 8 9 0 1
+-+-+-+-+-+-+-+-+-+-+-+-+-+-+-+-+-+-+-+-+-+-+-+-+-+-+-+-+-+-+-+-+
|U|          Message Type       |       Message Length          |
+-+-+-+-+-+-+-+-+-+-+-+-+-+-+-+-+-+-+-+-+-+-+-+-+-+-+-+-+-+-+-+-+
|                         Message ID                            |
+-+-+-+-+-+-+-+-+-+-+-+-+-+-+-+-+-+-+-+-+-+-+-+-+-+-+-+-+-+-+-+-+
|                                                               |
+                                                               +
|                     Mandatory Parameters                      |
+                                                               +
|                                                               |
+-+-+-+-+-+-+-+-+-+-+-+-+-+-+-+-+-+-+-+-+-+-+-+-+-+-+-+-+-+-+-+-+
|                                                               |
+                                                               +
|                     Optional Parameters                       |
+                                                               +
|                                                               |
+-+-+-+-+-+-+-+-+-+-+-+-+-+-+-+-+-+-+-+-+-+-+-+-+-+-+-+-+-+-+-+-+
```

Figure 4.9. *Generic format of LDP message*

– Message ID: this field inserts the identifier of the LDP message used by the receiver when issuing the notification message on a received message.

– Mandatory Parameters: this block inserts the required parameters of the message in the form of a TLV (Type Length Value) structure.

– Optional Parameters: this block inserts the optional parameters of the message in the form of a TLV structure.

4.1.5.3. *LDP messages*

Table 4.2 summarizes the parameters associated with the LDP messages.

The LSRs declare to adjacent routers by issuing a HELLO message that contains the following parameters:

– mandatory parameter *hello*. This parameter indicates the value of the timer Hello Time. This timer specifies the maximum waiting time a HELLO message from the adjacent LSR. The default value is 15 seconds in the case of a multicast transmission and 45 seconds in the case of a unicast transmission. If two routers advertise two different values, the lowest value is retained.

– optional parameter *IPv4 address*. This parameter contains the IP address of the router LSR that determines the active router LSR that initiates the TCP session.

– optional parameter *sequence number*. The increment of the sequence number initializes a new discovery step if the previous one cannot succeed.

Parameters	Messages	Mandatory/optional
Session	INITIALIZATION	Mandatory
address list	ADDRESS	Mandatory
	ADDRESS_WITHDRAW	Mandatory
FEC	LABEL_MAPPING	Mandatory
	LABEL_REQUEST	Mandatory
	LABEL_WITHDRAW	Mandatory
	LABEL_RELEASE	Mandatory
	LABEL_ABORT_REQUEST	Mandatory
label	LABEL_MAPPING	Mandatory
	LABEL_WITHDRAW	Optional
	LABEL_RELEASE	Optional
hop count	LABEL_MAPPING	Optional
	LABEL_REQUEST	Optional
path vector	LABEL_MAPPING	Optional
	LABEL_REQUEST	Optional
request message	LABEL_MAPPING	Optional
	LABEL_ABORT_REQUEST	Optional
status	NOTIFICATION	Optional
extended status	NOTIFICATION	Optional
returned PDU	NOTIFICATION	Optional
returned message	NOTIFICATION	Optional

Table 4.2. *LDP message parameters*

The INITIALIZATION message establishes an LDP session between two adjacent LSRs. When the TCP session is established, the active router sends the INITIALIZATION message that contains the following parameter:

– mandatory parameter *session*. This parameter determines the value of the keepalive time of the LDP session, the label distribution method (Unsolicited Downstream or Downstream On Demand), the maximum size of LDP PDU, an indication of the implementation of the loop detection and the maximum number of hops.

Passive LSR can respond with INITIALIZATION message or KEEPALIVE message. If the required parameter is not accepted by the passive router, it responds with the NOTIFICATION message.

LSRs use KEEPALIVE messages to maintain the LDP session. This message has no parameters. If no message LDP is not received at the end of keepalive time, LDP session is released.

ADDRESS message provides the router LSR the knowledge of IP addresses of the next hop LSR. It is transmitted after initialization of the LDP session or when a new address is configured. It contains the following parameters:

– mandatory parameter *address list*. This parameter determines the list of IP addresses of the downstream router interface.

ADDRESS WITHDRAW message allows LSR to remove the IP addresses of the next hop LSR. It contains the same parameter as the ADDRESS message.

LABEL_MAPPING message is used by the downstream LSR to indicate the upstream router the value of label associated with the IP address. This message is sent by the downstream router of its own (Unsolicited Downstream) or following a request from the upstream router (Downstream On Demand). The LABEL_MAPPING message includes the following parameters:

– mandatory parameter FEC. This parameter contains the IP address of the LSR for which a label is set.

– mandatory parameter *label*. This parameter contains the value of the label associated with the IP address.

– optional parameter *hop count*. This parameter contains the number of hops made by the message. The value is incremented by one for each crossed LSR. The limit value of the number of hops is determined by the parameter *session* of INITIALIZATION message.

– optional parameter *path vector*. This parameter contains the list of the IP addresses for the routers traversed. It can detect a loop at the transmission of LDP messages.

– optional parameter *request message*. This parameter contains the identifier of the request expressed by LABEL_REQUEST message.

LABEL_REQUEST message is used in the Downstream On Demand mode by the upstream router to request the value of a label associated with the IP address of the LSR. It contains the mandatory parameter FEC and the optional parameters *hop count* and *path vector*.

LABEL WITHDRAW message is used by the downstream router for the removal of a label. It contains the mandatory parameters FEC and *label*.

LABEL RELEASE message is sent by the upstream router to cancel the label issued by the downstream router. It contains the mandatory parameters FEC and *label*.

ABORT LABEL REQUEST message is used by the upstream router to cancel a LABEL_REQUEST query. It contains the mandatory parameters FEC and *request message*.

Routers use NOTIFICATION message indicating an error in the procedure. If the error is fatal, LDP session is released. NOTIFICATION message contains the following parameters:

– mandatory parameter *status*. This parameter provides an indication of the treatment of error (fatal error, error to be transferred) and the type of error detected.

– optional parameter *extended status*. This parameter provides additional indications on the type of error.

– optional parameter *returned* PDU. This parameter provides a copy of the PDU that caused the error.

– optional parameter *returned message*. This parameter provides a copy of the message that caused the error.

4.2. VPN function

4.2.1. *MPLS-VPN architecture*

The introduction of the VPN (Virtual Private Network) function in the MPLS network is obtained from the following two functions, located in the Edge LSR:

– the introduction of routing instances, called VRF (VPN Routing and Forwarding) associated with each interface of the Edge LSR, at the user side;

– the marking of labeled IP packets belonging to the same group of users. This marking, corresponding to a second MPLS label (VPN label), is effective between two Edge LSRs.

The components of MPLS-VPN network are the following (Figure 4.10):

– CE (Customer Edge): this is usually the router for the user of the MPLS-VPN network, which establishes adjacency with the router of the MPLS-VPN network (PE router). It does not participate in the labeling of IP packets.

– PE (Provider Edge): this is the new name of the Edge LSR. It implements the VPN feature in addition to the functions of the Edge LSR.

– P (Provider): this is the new name of the Core LSR. It has no knowledge of the VPN feature. Its operation is identical that of the Core LSR.

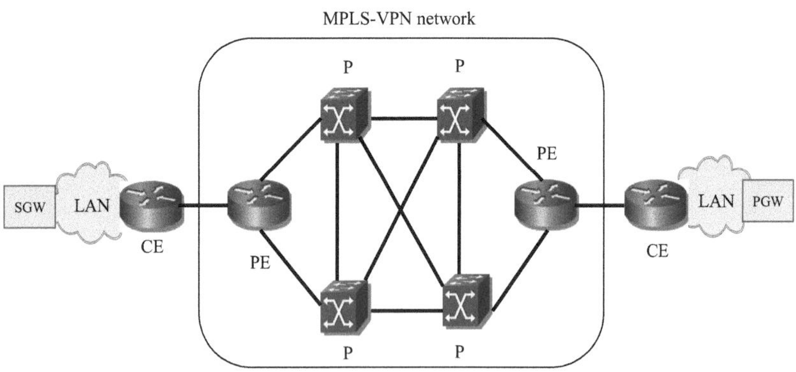

Figure 4.10. *Functional architecture of the MPLS-VPN network*

MPLS-VPN network can route packets exchanged between the entities of 4G mobile core network. The CE routers are gateways for the LAN (Local Area Network) to which these entities are connected.

The CE router sends the PE entity announcements related to local routes, and receives from the PE entity announcements related to remote routes. Announcements are made from standard routing protocols (RIPv2, OSPF, eBGP).

The PE entity maintains the routing instances VRF (Figure 4.11). A physical or logical interface, at the CE side, is associated with a VRF. Several interfaces, at the CE side, can be grouped on a single VRF.

The PE entity advertises to other PE entities, with VPN routes, built from routes provided by thee CE router. A VPN route is an aggregate of the IPv4 network address and a Route Distinguisher (RD). Announcements are made from MP-BGP (Multi-Protocol Border Gateway Protocol) routing protocol.

The P entity performs the switching of the LSP label in order to transfer IP packets between two PE entities. The switching is performed from LSP label that is different from the VPN label assigned to the marking of flows belonging to the same user group.

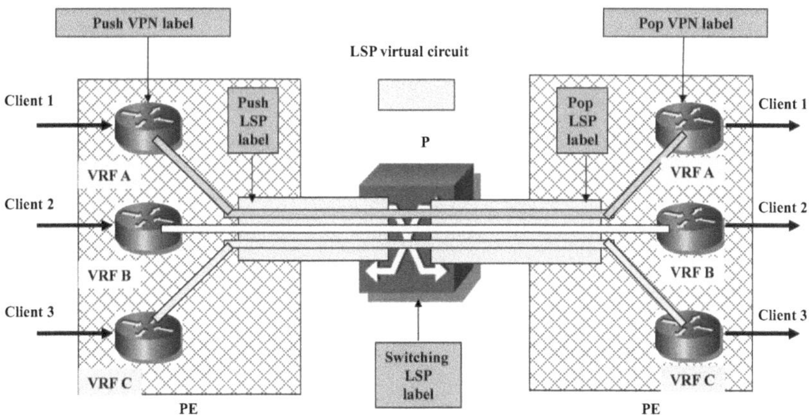

Figure 4.11. *VPN construction. For a color version of this figure, see www.iste.co.uk/perez/4gmobile.zip*

The label switching uses the principles described for the MPLS network:

– OSPF and IS-IS protocols allow to know the routes to reach the IP address of the PE entities.

– LDP allows to assign a label to that IP address.

These data belonging to the control plane are exchanged between adjacent entities (PE and P) of the MPLS-VPN network and enable the establishment of the virtual circuit LSP.

The distribution of VPN routes and the allocation of the VPN label, marking the IP packet belonging to the same group of users, are obtained from the MP-BGP routing protocol. These data are exchanged between the PE entities.

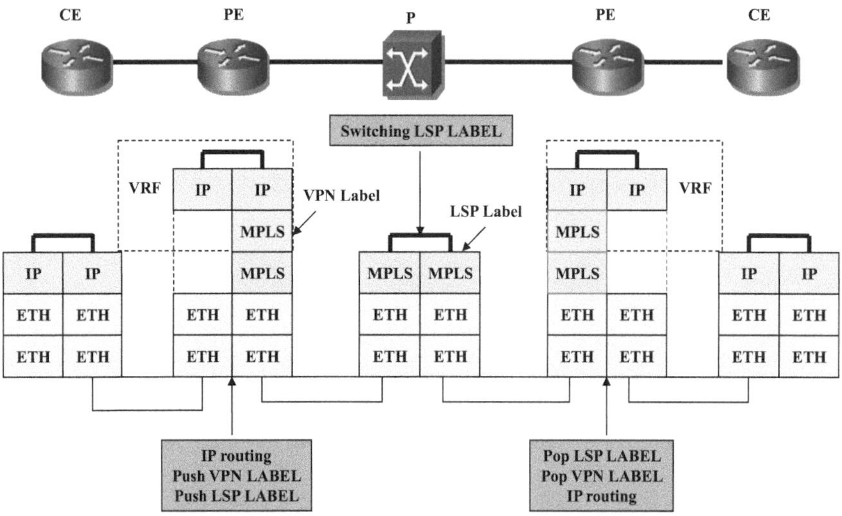

Figure 4.12. *Protocol architecture of the MPLS-VPN network*

Ingress PE entity performs routing of incoming IP packet and introduces two labels (Figure 4.12):

– the first label carries the VPN label inserted from the information provided by the routing table (RIB) dedicated to the VRF instance;

– the second label contains the LSP label used for label switching inserted from the information of the label switching table (LFIB).

P entity performs label switching by acting only on the LSP label.

Egress PE entity removes both labels and performs routing of the IP packet.

4.2.2. *Route distinguisher*

The PE entity must transfer the learned routes from CE router to other PE entities via the MP-BGP. It offered the opportunity to have an IP address overlap between customers belonging to different VPNs. It even allowed the customer to use private addresses. Typically each VPN can use the address range 10.X.X.X.

There are two types of routing table (RIB) at the PE entity: routing tables dedicated to each VRF and common routing tables. Routing protocols, especially MP-BGP, advise common routing tables to limit the number of instances.

The route distinguisher is obtained by adding an RD identifier. Each VRF instance of the same VPN can utilize the same RD. The RD identifier is not used to build a VPN but only to distinguish between identical routes.

The RD identifier is a 64-bit field. A VPNv4 route is the aggregation in an RD identifier and an IPv4 route. Its size is equal to 96 bits. The RD identifier is configured for each VRF instance. The RD identifier is constructed with two fields:

– the first 32-bit field may be the Autonomous System number or an IP address (that of PE entity, for example);

– the second 32-bit field inserting its value dedicated to the VRF.

4.2.3. *Route target*

The construction of VPN is based on the Route Target (RT). RT identifier structure is identical to that of RD identifier. The RT identifier defines how the routes will be inserted into the VRF instances. Each VRF instance is configured with two parameters:

– RT export: this parameter is transferred in the MP-BGP messages in association with the announced VPNv4 route. It constitutes an extended BGP community.

– RT import: this parameter is used to filter for inserting a VPNv4 route learned via the MP-BGP protocol. The VRF instance checks whether the received RT export has the same value as the RT import.

A mesh network is achieved by the combination of all sites in pairs. To exchange routes, each PE router establishes the same rule for import and export routes (Figure 4.13).

RT import = RT export.

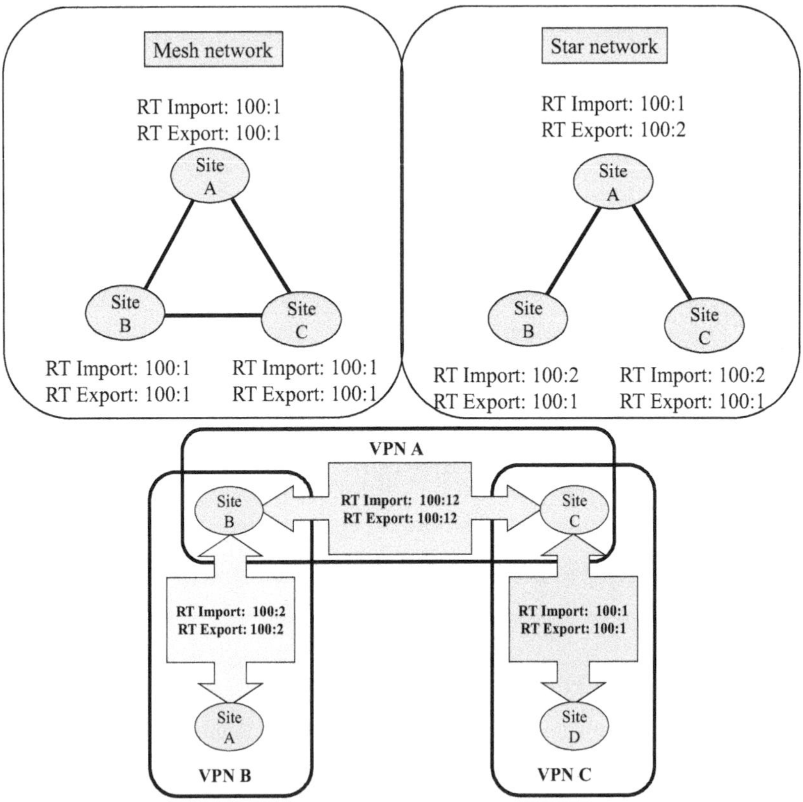

Figure 4.13. *VPN topologies*

A star network (hub and spoke) includes a central site and peripheral sites. The exchange of routes is carried out only between, firstly, the central site and, on the other hand, peripheral sites. The PE routers do not have the same rules of import and export routes (Figure 4.13).

RT import for central site = RT export for peripheral sites

RT export for central site = RT import for peripheral sites

RT export for peripheral sites ≠ RT import for peripheral sites

Communication between peripheral sites B and C is possible through the central site A, if the rules established by the central site A allow.

More complex topologies can be obtained, such as the possibility for a site to belong to several VPNs (Figure 4.13).

4.2.4. *Principles of operation*

4.2.4.1. *Control plane*

MPLS-VPN network implements data belonging to the control plane for the construction of the virtual circuit LSP between the PE routers and VPN.

In the following example, the PE1 address is a private address (10.1.1.1). When the addresses of the PE entities is known by all of the PE and P entities, the construction of the LSP virtual circuit operates as follows (Figure 4.14):

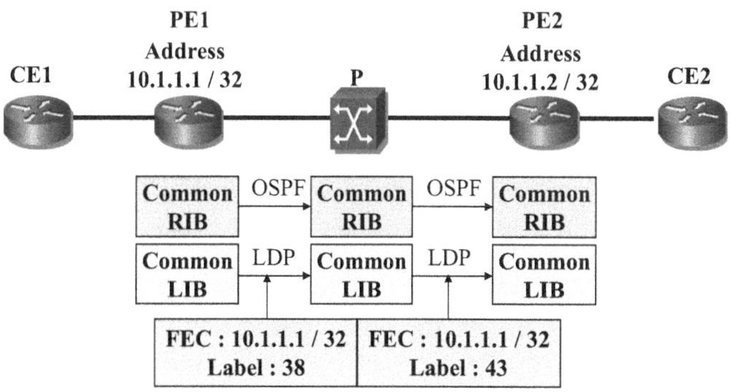

Figure 4.14. *Operation of the control plane Construction of the virtual circuit LSP*

– PE1 entity sends an LDP message to the P entity, indicating the value 38 attributed to the IP address (10.1.1.1/32) of the PE1 entity;

– P entity sends an LDP message to the PE2 entity, indicating the value 43 attributed to the IP address (10.1.1.1/32) of the PE1 entity.

VPN construction takes place in several stages (Figure 4.15):

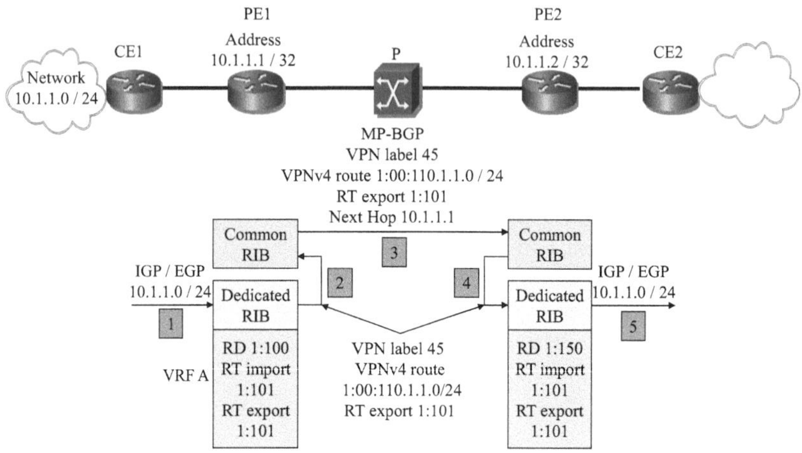

Figure 4.15. *Operation of the control plane Construction of the VPN*

– step 1: CE1 router performs a route announcement (10.1.1.0/24) to the VRF instance A of PE1 entity. It was caught knowingly identical addresses between, on the one hand, customers attached to CE1 routers, and, secondly, the entities of the MPLS-VPN network in order to show that there is no overlap. The VRF instance A updates its RIB table, transforms the IPv4 route into VPNv4 route by aggregating the RD identifier (1:100) and allocates the VPN label 45.

– step 2: VRF instance A transmits such information, and the RT export (1:101), to the common routing table RIB, which updates the common LFIB.

– step 3: PE1 entity transmits this information to the common routing table RIB of the PE2 entity, in the MP-BGP message that also contains the address of the next hop (10.1.1.1), that of the PE1 entity.

– step 4: On the basis of the RT export value (1:101), the common routing table transmits the MP-BGP announcement to the VRF instance A.

– step 5: VRF instance A removes the RD identifier of the VPNv4 route, updates its routing table RIB and transfers the address network (10.1.1.0/24) to the CE2 router.

The common routing table RIB of PE entities contains the address (10.1.1.1/32) of PE1 entity and the network address 1:100:10.1.1.0/24 attached to the CE1 router. The RD identifier (1:100) allows to remove the ambiguity.

The routing table RIB dedicated to a VRF instance does not know the IP addresses assigned to PE and P entity of the MPLS-VPN network.

4.2.4.2. *Traffic plane*

The transfer of data belonging to the traffic plane, between CE1 and CE2 routers, implements the following (Figure 4.16):

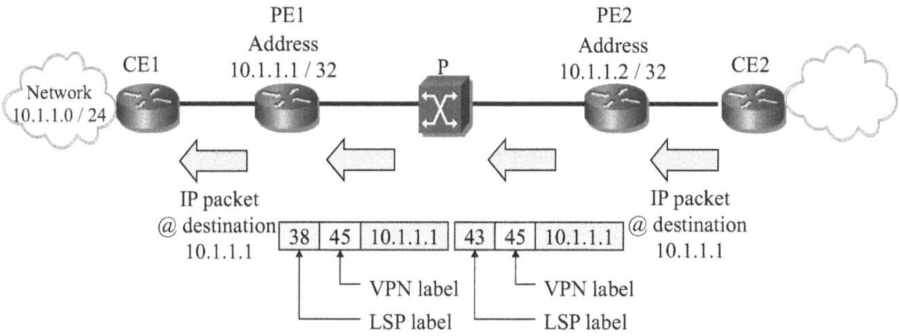

Figure 4.16. *Operation of the traffic plane*

– CE2 router transmits the PE2 entity an IP packet whose destination address is 10.1.1.1. This address is not that of the PE1 entity but that of a host belonging to the network 10.1.1.0/24.

– VRF instance A of the PE2 entity performs the VPN label with the value 45. The PE2 entity consults the common LFIB table to determine the destination address that of PE1 router (10.1.1.1). The common LFIB table assigns an LSP label (value 43) to reach the PE1 entity.

– P entity performs switching labels and replaces the value of the received label 43 by the value 38.

– PE1 entity removes the label 38 and uses the value of the VPN label (value 45) to determine the VRF instance. The VRF instance A transmits the IP packet to the CE2 router.

IP packet containing the destination address 10.1.1.1 is not addressed to the PE1 entity because this packet is labeled 45.

The routing table RIB dedicated to the VRF instance A shows for this destination address output to CE1 router. The address 10.1.1.1 of the PE1 entity is not known in the routing table RIB dedicated to the VRF instance A, but only at the common routing table RIB.

5

Transport Network VPLS Technology

5.1. Network architecture

5.1.1. *VPLS network*

Virtual Private LAN Service (VPLS) network is a network that connects the Evolved Node Base station (eNB) to the Local Area Network (LAN) that hosts Mobility Management Entity (MME) and Serving Gateway (SGW), entities of the 4G mobile core network (Figure 5.1).

The architecture of the VPLS network is based on that of the Multi-MPLS-VPN network (Figure 5.1):

– Provider Edge (PE) entity hosts Virtual Switch Instance (VSI), an Ethernet switch, instead of VPN Routing and Forwarding (VRF), an IP router;

– Provider (P) provides the same label switching function, regardless of the type of PE.

The architecture of the VPLS network is meshed, and Ethernet connectivity is ensured between any PE entities.

PE entity performs switching of the incoming Ethernet frame, from the CE router, and two labels (Figure 5.2):

– the first label VCL (Virtual Channel Label) bears the mark for the VPN (Virtual Private Network) group. This mark is equivalent to the VPN label of the MPLS-VPN network;

– the second label TL (Tunnel Label) contains the identifier for the label witching carried by P entities and the constitution of the tunnel. This mark is equivalent to the LSP label of the MPLS-VPN network.

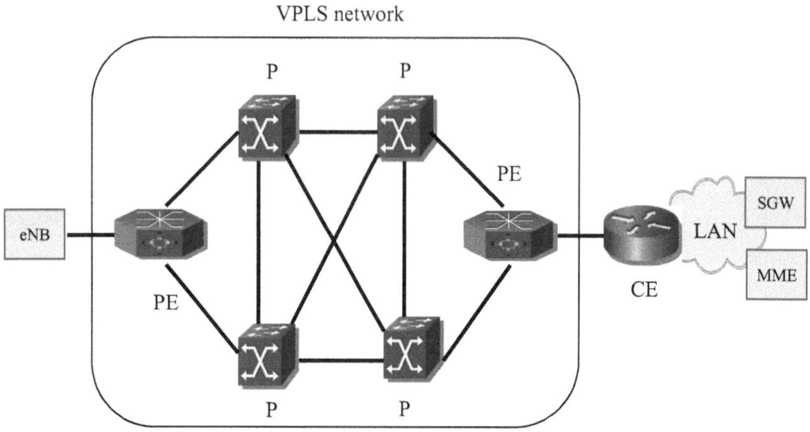

Figure 5.1. *Functional architecture of the VPLS network*

P entity performs switching labels by acting only on the second label TL (Figure 5.2).

Egress PE entity removes both labels, performs switching of the Ethernet frame and transfers the frame to the CE router (Figure 5.2).

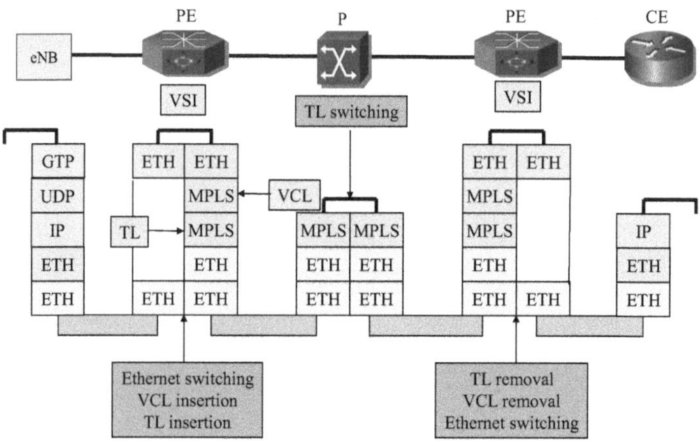

Figure 5.2. *Protocol architecture of the VPLS network*

TL label switching uses the principles described for the MPLS network:

– OSPF (Open Shortest Path First) or IS-IS (Intermediate System to Intermediate System) allow to know the routes to reach PE entities;

– LDP (Label Distribution Protocol) allows to assign the label TL to the PE entity. The TL label is carried by a part of the EoMPLS (Ethernet over MPLS) header, identical to an MPLS header and interpreted as such by the P entity. The allocation of the TL label is performed by exchanging LDP messages between adjacent PE or P entities (Figure 5.3).

VCL label is also carried by the EoMPLS header and can substitute, in the case of a Q-in-Q interface type, S-VLAN (Virtual Local Area Network Service) mark, which is deleted.

The establishment of the VCL label between two virtual switches VSI is performed either manually or dynamically from LDP or MP-BGP (Multi-Protocol Border Gateway Protocol) messages exchanged between PE entities (Figure 5.3).

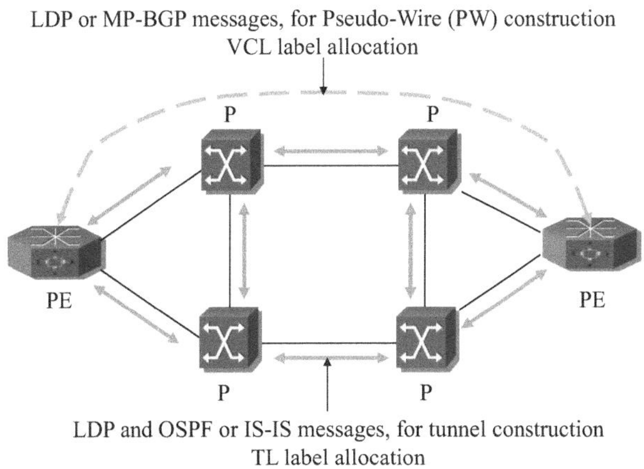

Figure 5.3. *OSPF, LDP and MP-BGP message exchanges*

The virtual switch VSI has two types of interfaces:

– PW (Pseudo-Wire) interface, at VPLS network side, identified by the VCL label;

– AC (Attachment Circuit) interface, that is Ethernet type, at client side whose header may not contain any VLAN tag or VLAN tag (802.1Q Ethernet) or two Ethernet marks (Q-in-Q), C-VLAN (Customer VLAN) and S-VLAN.

The correspondence between, on the one hand, the physical access or the VLAN tag number of the AC interface, and, secondly, the VSI instance is configured manually.

5.1.2. *H-VPLS network*

H-VPLS (Hierarchical VPLS) architecture can extend the VPLS network, the PE equipment being deployed at two levels (Figure 5.4):

– N-PE (Network-facing PE). The interconnection of N-PE entities is a mesh of virtual circuits or tunnels; N-PE entities being connected in pairs. Entities from the 4G mobile core network are connected to the N-PE entity.

– U-PE (User-facing PE). The interconnection between the U-PE and N-PE equipment is a star of tunnels; U-PE entities are only connected to a single N-PE. This arrangement optimizes the number of tunnels. The eNB entities are connected to the U-PE entity.

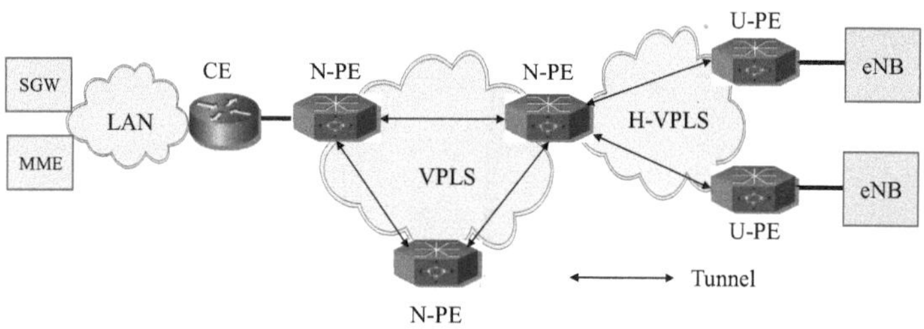

Figure 5.4. *Functional architecture of the H-VPLS network*

5.2. MAC address management

The switching table of the VSI instance is populated by learning, like an Ethernet switch. The table shows, for each MAC address, the output port (AC or PW interfaces).

On the other side, the operating mode of the virtual switch is different with regard to the broadcast or multicast frame.

In an Ethernet switch network, Spanning Tree Protocol (STP) is used to establish a spanning tree and to avoid flooding the network with broadcast frames, broadcast, multicast or unicast if the switching table is empty.

A different mechanism, split horizon, is used in the network of virtual switches (Figure 5.5).

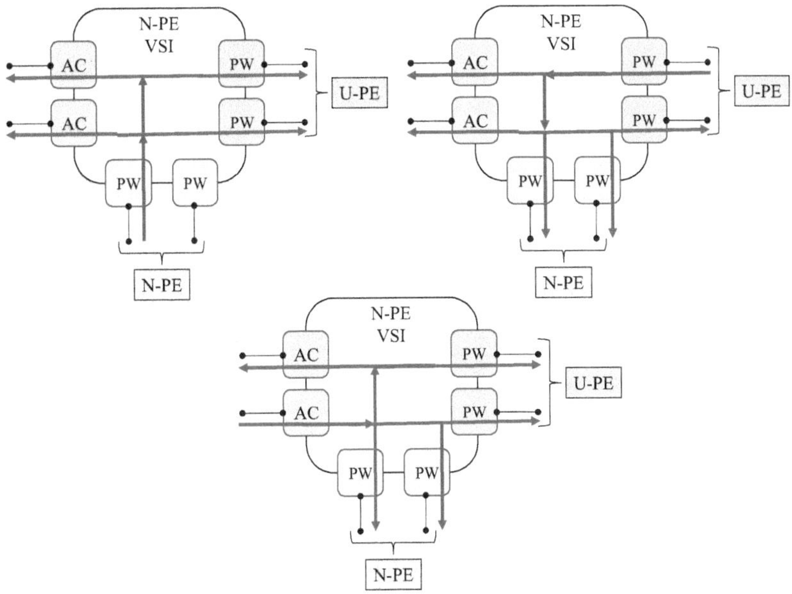

Figure 5.5. *Broadcasting the frames*

When the VSI instance of an N-PE entity receives a frame on a PW access from another N-PE entities, it prohibits the diffusion to other PW accesses associated with N-PE entities and allows diffusion to AC accesses and PW accesses associated with U-PE entity.

When the VSI instance of an N-PE entity receives a frame on a PW access from a U-PE entity, it allows for diffusion to AC accesses, to PW accesses associated with N-PE entities and to other PW accesses associated with U-PE entity.

When the VSI instance of an N-PE entity receives a frame on an AC access, it allows for diffusion to other AC accesses and to PW accesses associated with N-PE or U-PE entities.

PE entity that detected a change in the topology must inform other PE entities so they withdraw the impacted MAC addresses of the switching table of the VSI instances.

If the topology change is on AC or PW access, N-PE entity broadcasts the withdrawal of the MAC addresses on the PW accesses associated with N-PE and U-PE entities.

The withdrawal information of the MAC addresses received by an N-PE entity is not broadcast to other N-PE entities.

5.3. EoMPLS header

EoMPLS header encapsulates an Ethernet frame whose header is stripped of the following fields: Preamble, Start of Frame Delimiter (SFD) and Frame Check Sequence (FCS).

EoMPLS header consists of three 32-bit words (Figure 5.6):

– the first two words correspond to a conventional MPLS header;

– the third word is a control word.

The first word contains the value assigned dynamically to the TL label by the LDP protocol. The EXP field can be used to manage priorities. Bit S (stack) is set to ZERO to indicate that the bottom of the label stack is not achieved. The value of the TTL field is set by the PE entity performing the insertion of TL label. This field is decremented by one for each crossing P entity.

The second word contains the value assigned to the label VCL. The EXP field is not used. Bit S (stack) is set to ONE to indicate that the bottom of the label stack is reached. The value of the TTL field is set by the PE entity to 2. This value is decremented by the PE entity performing the removal of the VCL label.

The third word contains a 4-bit field to ZERO. The Sequence Number field, whose use is optional, is used to restore output Ethernet frames in sequence and to remove duplicate frames.

```
 0                   1                   2                   3
 0 1 2 3 4 5 6 7 8 9 0 1 2 3 4 5 6 7 8 9 0 1 2 3 4 5 6 7 8 9 0 1
+-+-+-+-+-+-+-+-+-+-+-+-+-+-+-+-+-+-+-+-+-+-+-+-+-+-+-+-+-+-+-+-+
|              Tunnel LABEL             | Exp |0|    TTL=255    |
+-+-+-+-+-+-+-+-+-+-+-+-+-+-+-+-+-+-+-+-+-+-+-+-+-+-+-+-+-+-+-+-+
|        Virtual Channel LABEL          | Exp |1|    TTL=2      |
+-+-+-+-+-+-+-+-+-+-+-+-+-+-+-+-+-+-+-+-+-+-+-+-+-+-+-+-+-+-+-+-+
|0 0 0 0|      Reserved         |        Sequence Number        |
+-+-+-+-+-+-+-+-+-+-+-+-+-+-+-+-+-+-+-+-+-+-+-+-+-+-+-+-+-+-+-+-+
```

Figure 5.6. *EoMPLS header format*

5.4. LDP protocol

LDP messages are exchanged between PE entities for the establishment of the VCL label associated with the implementation of the pseudo-wire (PW).

The evolution made to the LDP protocol are related to TLV (Type Length Value) parameters of some messages. The setting process of the pseudo-wire between PE entities is the following:

– PW interface is configured at each PE entity and its correspondence with the VSI instance;

– each PE entity initiates a LDP session with remote PE entity. The PE entities exchange HELLO discovery messages and INITIALIZATION message for the establishment of the LDP session;

– each PE entity allocates a VCL label and an identifier to the pseudo-wire. The VCL label is positioned in the *label* parameter and the identifier of the pseudo-wire in the FEC (Forwarding Equivalence Class) parameter. These two parameters are transmitted in a LABEL_MAPPING message.

5.4.1. *FEC parameter*

FEC parameter allows to create the link between the identifier of pseudo-wire and the VCL label. The structure of the FEC parameter contains the following fields (Figure 5.7):

– PWid: this field identifies the type of FEC. Its value is 80 in hexadecimal for this type of FEC described.

– C-Bit: this bit indicates whether the control word of the EoMPLS header is used (bit to ONE) or not (bit to ZERO).

– PW Type: this field indicates the type of the pseudo-wire. It takes the following values in hexadecimal:

- 0004, if the AC interface is an Ethernet interface containing a VLAN tag;

- 0005, if the AC interface is an Ethernet interface without VLAN tag;

– PW Info Length: this field indicates the size of the PW ID field and the subparameter TLV Interface Parameters.

– Group ID: this field encodes a value assigned to a set of pseudo-wires. The withdrawal of label is obtained from LABEL_WITHDRAW message. You can optionally remove a set of labels for a PW group identified by the Group ID field.

– PW ID: this field encodes the value of the identifier of the pseudo-wire. For this type of FEC, the same identifier (PW ID) is used at both ends of the pseudo-wire.

– Interface Parameters: this subparameter has a TLV structure providing information on the AC interface, such as the MTU size of the load of the Ethernet frame.

```
 0                   1                   2                   3
 0 1 2 3 4 5 6 7 8 9 0 1 2 3 4 5 6 7 8 9 0 1 2 3 4 5 6 7 8 9 0 1
+-+-+-+-+-+-+-+-+-+-+-+-+-+-+-+-+-+-+-+-+-+-+-+-+-+-+-+-+-+-+-+-+
|    PWid       |C|             PW Type         |PW info Length |
+-+-+-+-+-+-+-+-+-+-+-+-+-+-+-+-+-+-+-+-+-+-+-+-+-+-+-+-+-+-+-+-+
|                            Group ID                           |
+-+-+-+-+-+-+-+-+-+-+-+-+-+-+-+-+-+-+-+-+-+-+-+-+-+-+-+-+-+-+-+-+
|                             PW ID                             |
+-+-+-+-+-+-+-+-+-+-+-+-+-+-+-+-+-+-+-+-+-+-+-+-+-+-+-+-+-+-+-+-+
|                                                               |
//                 Interface Parameters TLV                    //
|                                                               |
+-+-+-+-+-+-+-+-+-+-+-+-+-+-+-+-+-+-+-+-+-+-+-+-+-+-+-+-+-+-+-+-+
```

Figure 5.7. *FEC parameter format*

5.4.2. *PW Status parameter*

PE entities can use the PW Status parameter carried in the NOTIFICATION message to notify the remote PE entity of the state of pseudo-wire. The negotiation procedure for the use of PW Status parameter is as follows:

– PE entity that supports this feature includes the PW Status parameter in the LABEL_MAPPING message;

– if the remote PE entity does not support this feature, it responds with the LABEL_RELEASE message, indicating the reason for releasing, PW Label Withdraw Method Status Not Supported.

In case the LABEL MAPPING message does not contain PW Status parameter, the function is not used.

5.4.3 *Address Withdraw message*

The Address Withdraw message has been introduced to remove MAC addresses of the switching table of the VSI instances during a network reconfiguration.

The Address Withdraw message contains the FEC parameter to identify the VSI instances affected by the withdrawal of MAC addresses and the MAC List parameter to list the MAC addresses to be removed.

5.5. MP-BGP protocol

The MP-BGP protocol is exchanged between the PE entities for the discovery of VSI instances and for the exchange of VCL label associated with the pseudo-wire.

The use of Route Reflector (RR) limits the number of BGP sessions between PE entities. Each PE entity only establishes a session with the route reflector, which in turn transfers the information to other PE entities.

When a new PE entity is added, it informs the other PE entities via the route reflector, which simplifies the configuration of other PE entities.

The MP-BGP protocol has introduced a new format for Network Layer Reachability Information (NLRI) transmitted in the UPDATE message containing the following parameters:

– Route Distinguisher (RD);

– VPLS Endpoint Identifier (VE-ID);

– VPLS endpoint Block Offset (VBO) of the first value of the VCL label used;

– VPLS endpoint Block Size (VBS) of the block of the VCL labels;

– Label Base (LB).

Route distinguisher (RD) identifies the VPLS domain. The route distinguisher is a concatenation of the Autonomous System (AS) number and the VPN identifier group.

Route distinguisher associated with the VSI instance number identifies the VSI instance.

VE-ID parameter identifies the PE entity that transmits the message BGP.

Local PE entity must announce at least one block of labels. The label block, whose size is specified by the parameter VBS, is a contiguous set of labels, in which the remote PE entities should select the value of the label.

The label base LB contains the first value in the list of VCL labels VCL. The last value of VCL labels is equal to (LB + VBS – 1).

The VBO offset is used when more blocks must be created, and the value of the VBO offset is calculated from the following formula:

$$VBO = ROUND(VE\text{-}ID/VBS) \times VBS$$

Example: VE-ID = 20 and VBS = 8, VBO = ROUND(20/8) × 8 = 16

Remote PE entity checks if its VE-ID is in the range of the configuration of labels from the following formula:

$$VBO <= VE\text{-}ID < VBO + VBS$$

Example: VE-ID of the remote PE entity is equal to 19 (between 16 and 23).

The value of the VCL label used to forward traffic to the local PE entity is equal to (LB + VE-ID – VBO).

Example: VCL = LB + 23 – 16 = LB + 7

The structure TLV Relearn Sequence Number (RSN) has been introduced to remove MAC addresses from the switching table of the VSI instances during a network reconfiguration.

When the RSN structure is received, its sequence number is compared with the previously received. If the new sequence number is greater than the former, the PE entity clears the MAC addresses of the corresponding VSI instance.

The extended community Layer2 Info Extended Community is associated with the NLRI announcement and contains encapsulation information:

– the field Extended Community identifies the type of extended community and takes the value 800A in hexadecimal;

– the field Encaps Type is set to 19 for VPLS technology;

– the field Control Flags contains the following information:

- the bit C indicates whether the control word of the EoMPLS header must be present (bit to ONE) or not (bit to ZERO),

- the S bit indicates whether the delivery of Ethernet frames is made in sequence (bit to ONE) or not (bit to ZERO);

– the field Layer-2 MTU indicates the maximum load size of the Ethernet frame used for the PW.

The Route Target (RT) is also associated with the NLRI announcement and its usage is identical to that described for the MPLS-VPN technology.

6

Transport Network OTN Technology

6.1. OTN architecture

The Optical Transport Network (OTN) is a transmission system on optical fiber. The solution based on Wavelength-Division Multiplexing (WDM) and Time-Division Multiplexing (TDM) allows to use the bandwidth of the optical fiber.

The optical transport network consists of the following equipment (Figure 6.1):

– OLT (Optical Line Terminal), used for establishment of point-to-point link or linear topology;

– OADM (Optical Add Drop Multiplexer), used for establishment of linear or ring topology;

– OXC (Optical Cross-Connect), used for establishment of mesh network.

The optical transport network (OTN) allows us to connect the PE (Provider Edge) and P (Provider) entities of MPLS-VPN (Multi-Protocol Label Switching - Virtual Private Network) or VPLS (Virtual Private LAN Service) networks.

The linear or ring networks are suitable for collecting different U-PE (User-facing PE) entities to the N-PE (Network-facing PE) entity of the H-VPLS (Hierarchical VPLS) network.

The mesh network is suitable for connecting entities of the MPLS-VPN or VPLS networks, connecting entities from the 4G mobile core network.

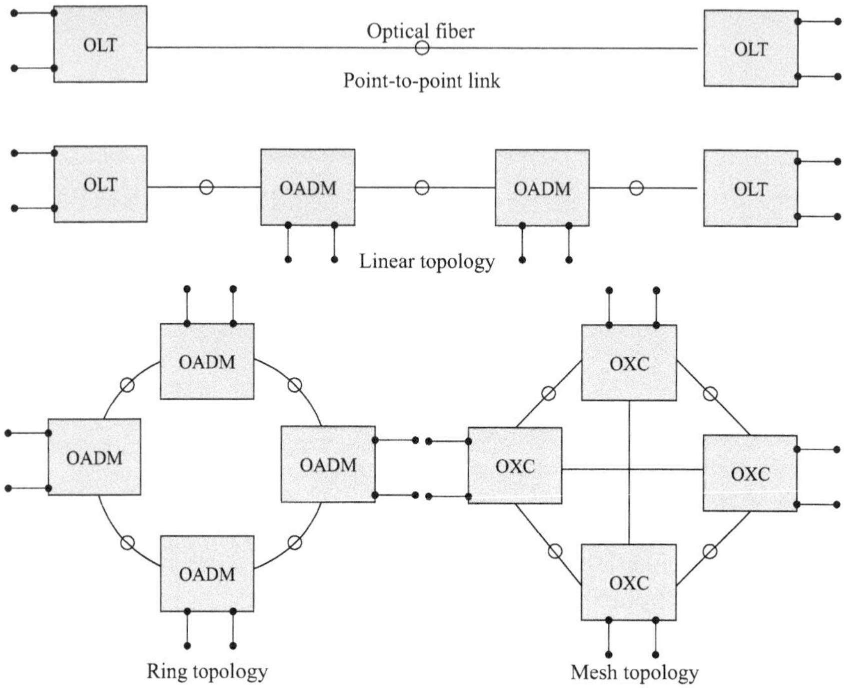

Figure 6.1. *OTN architecture*

The frequency grid defined for the Dense WDM (DWDM) system has various channel intervals, ranging from 12.5 to 100 GHz. Allowed frequencies in THz (Tera Hertz) are defined by the following formulas:

$$193.1 + n \times 0.0125/193.1 + n \times 0.025/193.1 + n \times 0.05/193.1 + n \times 0.1$$

where n is a positive integer, negative or zero.

The center frequency 193.1 THz corresponds to the wavelength of 1,552.52 nm. The step of 100 GHz corresponds approximately to a spacing of 0.8 nm. The wavelength band used is between 1,530 and 1,565 nm (C-band) and between 1,565 and 1,625 nm (L-band).

The nominal center wavelengths of the Coarse WDM (CWDM) system, whose values are between 1,271 nm and 1,611 nm, are spaced of 20 nm.

The WDM system uses the single-mode fiber G.652 or G.655 as a transmission medium.

The optical fiber G.652 is set in the O band (area between 1,260 and 1,360 nm) for which the chromatic dispersion is zero at 1,310 nm. It is also usable in the 1,500 nm window.

Four categories of G.652 fibers are based on the value of the Polarization Modal Dispersion (PMD) and the attenuation in the band E (area between 1,360 and 1,460 nm):

– G.652A: PMD = 0.5 ps × $km^{1/2}$, band E not usable;

– G.652B: PMD = 0.2 ps × $km^{1/2}$, band E not usable;

– G.652C: PMD = 0.5 ps × $km^{1/2}$, band E usable;

– G.652D: PMD = 0.2 ps × $km^{1/2}$, band E usable.

The optical fiber G.655 is defined in the 1,500 nm window. It is characterized by the lowest coefficient of chromatic dispersion than G.652 fiber, but the value must be greater than zero to minimize the increase in non-linear effects that can be particularly harmful in the WDM system.

The optical fiber G.655 C is originally intended to be used for wavelengths in the C band (area between 1,530 nm and 1,565 nm).

The optical fibers G.655D and G.655E define the chromatic dispersion characteristics in the S band (area between 1,460 and 1,530 nm) and L band (area between 1,565 nm and 1,625 nm).

6.1.1. *OLT equipment*

OLT equipment includes the following features (Figure 6.2):

– time-division multiplexing for tributary interfaces in order to form a digital signal whose rate can take several values (2.5, 10, 40 or 100 Gbps);

– transposing this signal only for the transmission in a specific wavelength using a transponder. This function corresponds to the OCh (Optical Channel) module;

– dense wavelength-division multiplexing (DWDM) characterized by a smaller channel separation than coarse wavelength-division multiplexing (CWDM). This function corresponds to the OMS (Optical Multiplex Section) module;

– Optical Amplifier (OA) of the transmitted signal (power amplifier) and the received signal (preamplifier) and optionally the Dispersion Compensation Module (DCM). This function corresponds to the OTS (Optical Transmission Section) module.

The optical modules OCh, OMS and OTS have their own over-head (OH) for optical control carried in the Optical Supervisory Channel (OSC) on a particular wavelength.

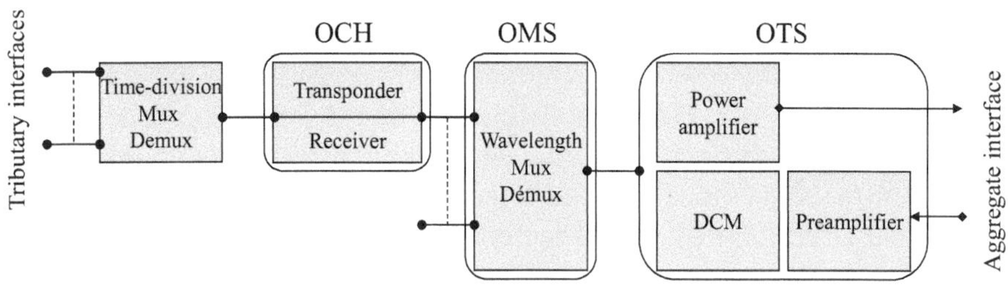

Figure 6.2. *OLT equipment structure*

6.1.2. *Regenerator equipment*

OLT equipment allows us to establish a point-to-point link. Transmission between 2 OLT equipment may require the implementetion of repeaters/regenerators whose function corresponds to 2 OTS modules head to tail.

3R regenerator performs in the optical domain the following functions: Re-shaping, Re-amplification and Re-timing.

Re-shaping is to correct distortions related to imperfections of the optical fiber. Re-amplification is designed to regenerate the signal to a level compatible with the transmission. Re-timing allows to regenerate the clock and to filter the signal jitter.

2R regenerator only in the optical domain realized the Re-shaping and the Re-amplification, which avoids converting the optical signal into an electrical signal.

6.1.3. *OADM equipment*

OADM equipment performs the wavelength transposition between aggregate interfaces and the insertion and extraction of wavelengths between the aggregate and tributary interfaces (Figure 6.3).

The insertion function is to add a wavelength in the optical multiplex, the latter from a tributary interface. The extraction function is to remove a wavelength of the optical multiplex in order to transmit it on a tributary interface.

The wavelength cross-connection function is performed by the Matrix Connection (MC).

The constitution of the aggregate signal is obtained from the optical modules OMS and OTS.

Access to tributary interfaces is done by the optical module OCh and the time-division multiplexing of tributary interfaces.

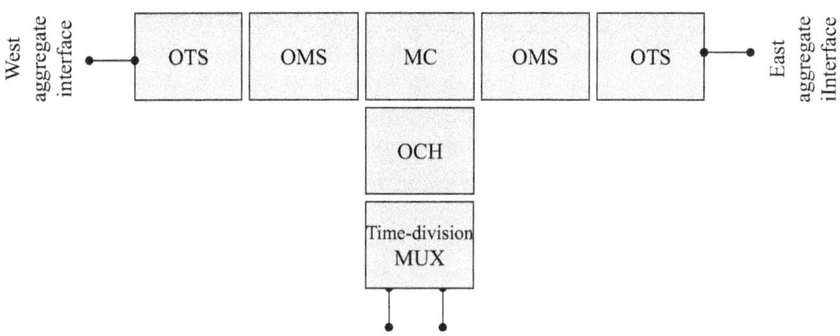

Figure 6.3. *OADM equipment structure*

6.1.4. *OXC equipment*

OXC equipment is an extension of the OADM equipment in that it has a larger number of aggregate interfaces (Figure 6.4).

Wavelength cross-connection consists of transferring a wavelength from the optical multiplex of an aggregate interface to the optical multiplex of another aggregate interface. This function is performed by the MC module.

The MC module also implements the wavelength conversion function for solving blocking status when optical channels at different inputs have the same wavelength.

OXC equipment can also provide access to tributary interfaces by the optical module OCh and time-division multiplexing.

6.2. Time-division multiplexing structure

Tributary Ethernet signal, from P or PE equipment of MPLS-VPN or VPLS network's, is mapped into the OPU (Optical channel Payload Unit) frame by adjusting the rate of the tributary interface to that of the ODU frame. The Over-Head (OH) of the OPU frame contains information that supports this adaptation mechanism (Figure 6.5).

OPU frame is mapped into ODU (Optical Channel Data Unit) frame whose over-head enables end-to-end supervision and the tandem connection monitoring of several OTN networks (Figure 6.5).

ODU frame is mapped into OTU (Optical channel Transport Unit) frame whose over-head allows for the synchronization of the OTU frame, the control of the time-division multiplexing section and the Forward Error Correction (FEC) (Figure 6.5).

OTU frame is carried by the optical channel OCh to which a wavelength is assigned in the optical multiplex.

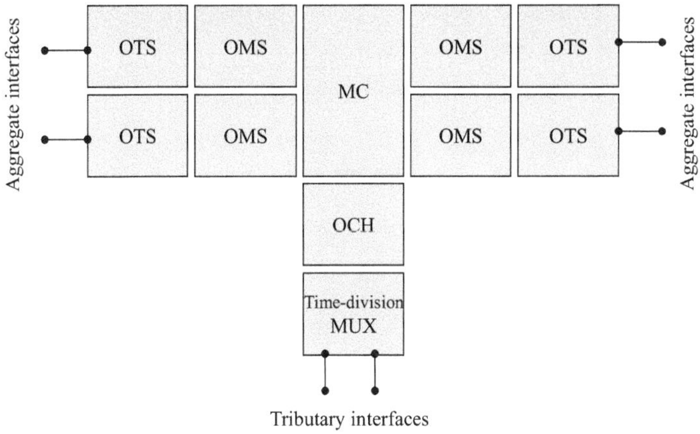

Figure 6.4. *OXC equipment structure*

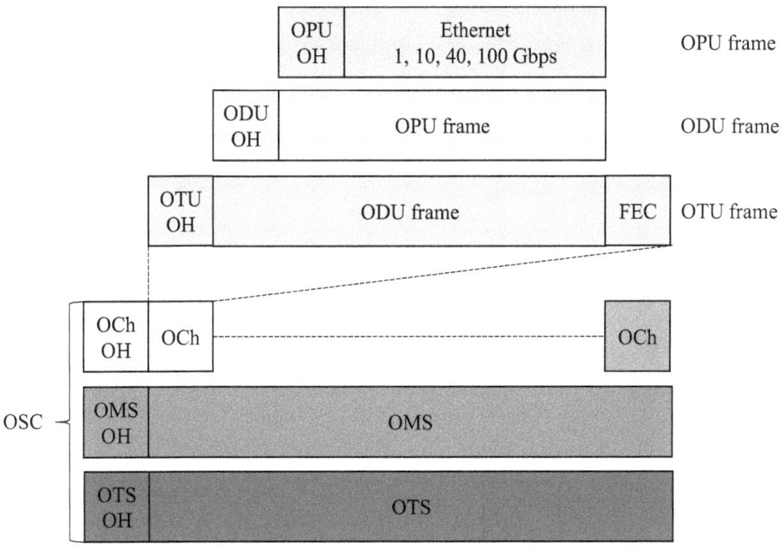

Figure 6.5. *General multiplexing structure*

OTU1 frame has a rate of 2.666 Gbps. The payload of the OTU1 frame is the ODU1 frame. The payload of the ODU1 frame is the OPU1 frame whose rate is adapted to 1 Gbps Ethernet interface (Figure 6.6).

The payload of the ODU1 frame also allows to multiplex 2 ODU0 frames. The payload of the ODU0 frame is the OPU0 frame whose rate is adapted to 1 Gbps Ethernet interface (Figure 6.6).

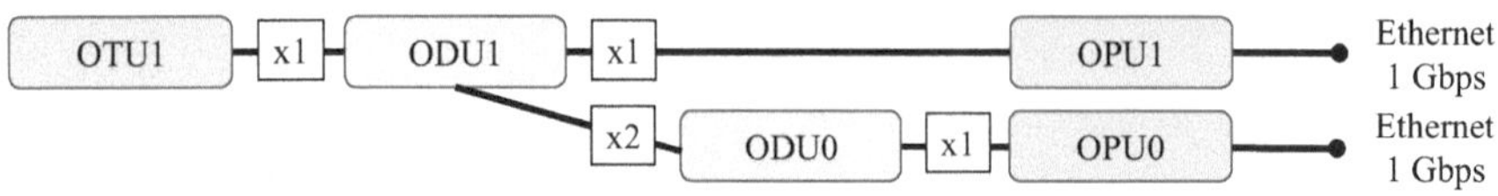

Figure 6.6. *OTU1 frame multiplexing structure*

OTU2 frame has a rate of 10.709 Gbps. The payload of the OTU2 frame is the ODU2 frame. The payload of the ODU2 frame is the OPU2 frame whose rate is adapted to 10 Gbps Ethernet interface (Figure 6.7).

The payload of the ODU2 frame also multiplexes the following frames:

– 4 ODU1 frames, adapted to 1 Gbps Ethernet interface;

– 8 ODU0 frames, adapted to 1 Gbps Ethernet interface.

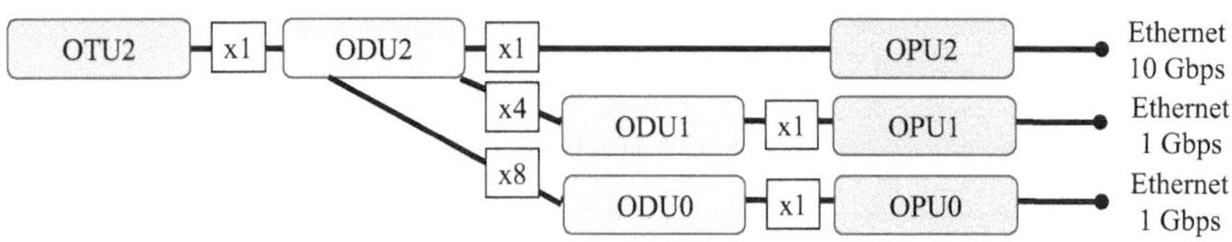

Figure 6.7. *OTU2 frame multiplexing structure*

OTU2e frame has a rate of 11.095 Gbps. The payload of the OTU2e frame is the ODU2e frame. The payload of the ODU2e frame is the OPU2e frame whose rate is adapted to 10 Gbps Ethernet interface (Figure 6.8).

Figure 6.8. *OTU2e frame multiplexing structure*

OTU3 frame has a rate of 43.018 Gbps. The payload of the OTU3 frame is the ODU3 frame. The payload of the ODU3 frame is the OPU3 frame whose rate is adapted to 40 Gbps Ethernet interface (Figure 6.9).

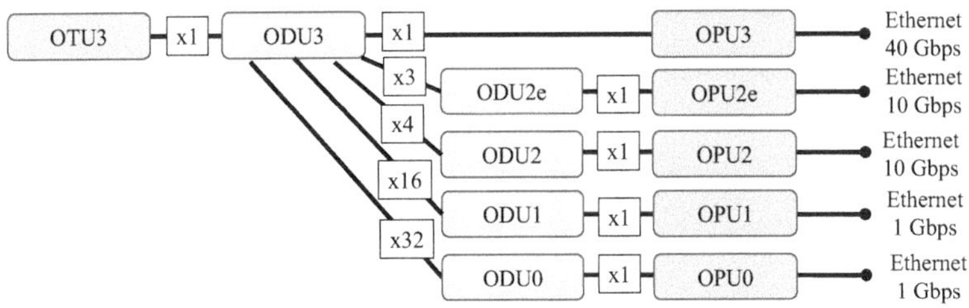

Figure 6.9. *OTU3 frame multiplexing structure*

The payload of the ODU3 frame also multiplexes the following frames:

– 3 ODU2e frames, adapted to 10 Gbps Ethernet interface;

– 4 ODU2 frames, adapted to 10 Gbps Ethernet interface;

– 16 ODU1 frames, adapted to 1 Gbps Ethernet interface;

– 32 ODU0 frames, adapted to 1 Gbps Ethernet interface.

OTU3e frame has a rate of 44.58 Gbps. The payload of the OTU3e frame is the ODU3e frame. The payload of the ODU3e frame is a multiplex of four ODU2e frames (Figure 6.10).

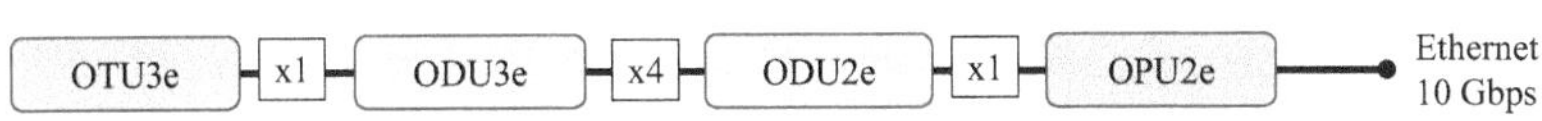

Figure 6.10. *OTU3e frame multiplexing structure*

OTU4 frame has a rate of 111.809 Gbps. The payload of the OTU4 frame is the ODU4 frame. The payload of the ODU4 frame is the OPU4 frame whose rate is adapted to 100 Gbps Ethernet interface (Figure 6.11).

The payload of the ODU4 frame also multiplexes the following frames:

– 2 ODU3 frames, adapted to 40-Gbps Ethernet interface;

– 10 ODU2e frames, adapted to 10-Gbps Ethernet interface;

– 10 ODU2 frames, adapted to 10-Gbps Ethernet interface;

– 40 ODU1 frames, adapted to 1-Gbps Ethernet interface;

– 80 ODU0 frames, adapted to 1-Gbps Ethernet interface.

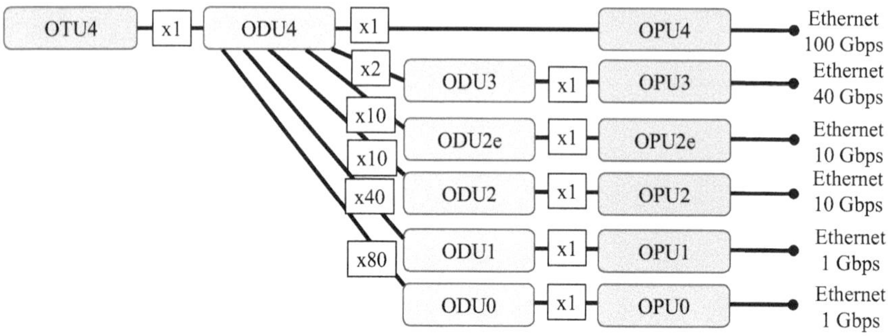

Figure 6.11. *OTU4 frame multiplexing structure*

6.3. Over-head structure

6.3.1. *OTS module over-head structure*

The over-head of the optical module OTS contains the following information:

– a Transmitted Trace Identifier (TTI);

– a Backward Defect Indication for the Payload (BDI-P) over an OTS section;

– a Backward Defect Indication for the Over-Head (BDI-O) over an OTS section;

– a Payload Missing Indication (PMI) for the OTS section.

6.3.2. *OMS module over-head structure*

The over-head of the optical module OMS contains the following information:

– a Forward Defect Indication for the Payload (FDI-P) over an OMS section;

– a Forward Defect Indication for the Over-Head (FDI-O) over an OMS section;

– a Backward Defect Indication for the Payload (BDI-P), over an OMS section;

– a Backward Defect Indication for the Over-Head (BDI-O) over an OMS section;

– a Payload Missing Indication (PMI) for the OMS section.

6.3.3. *OCh module over-head structure*

The over-head of the optical module OCh contains the following information:

– a Forward Defect Indication for the Payload (FDI-P), over an OCh section;

– a Forward Defect Indication for the Over-Head (FDI-O) over an OCh section;

– an Open Connection Indication (OCI). The signal is sent downstream. The detection subsequent to the loss of the OCh signal can then be associated with an open matrix.

6.3.4. *OTU frame over-head structure*

OTU frame consists of 4 rows and 4,080 columns of bytes, and includes the ODU frame completed by the following fields (Figure 6.12):

– the error correction (FEC) located in rows 1–4 and columns 3,825 to 4,080;

– the header OTU OH of the OTU frame located in row 1 and columns 8–14;

– the Frame Alignment Signal (FAS) and Multi-Frame Alignment Signal (MFAS) located in row 1 and columns 1–7.

To prevent the presence of long bit series in ZERO or ONE, an appropriate bit sequence is produced by the use of a scrambler, which is the generator polynomial $1 + x + {}^{3}x + x^{12} + x^{16}$. The frame alignment bytes are not to be scrambled. FAS field consists of 3 bytes OA1 (1111 0110) and 3 bytes OA2 (0010 1000) located in columns 1–6 of row 1 (Figure 6.12).

MFAS field, located in column 7 of row 1, allows the alignment of the multiframe (Figure 6.12). The value of the MFAS byte is incremented for each frame and allows the use of a multiframe consisting of 256 frames. The multiframe alignment is used by some of the fields of the headers spread over several frames.

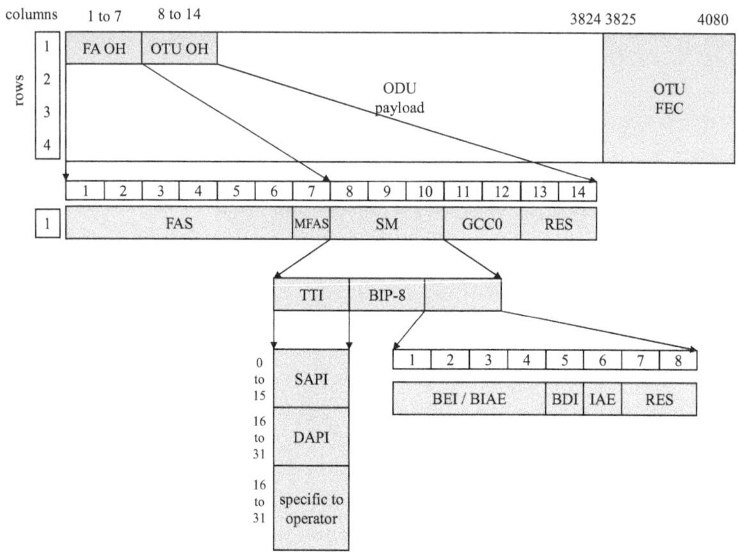

Figure 6.12. *OTU frame overhead structure*

SM (Monitoring Section) field, located in columns 8–10 of the row 1, contains the following subfields (Figure 6.12):

– a Transmitted Trace Identifier (TTI);

– a Bit Interleaved Parity (BIP-8) calculated on the bits in the DPU area (columns 15–3,824) of the frame i and inserted in frame i + 2;

– a Backward Defect Indication (BDI). BDI flag is set to ONE to indicate the presence of a defect;

– a Backward Error Indication (BEI) and a Backward Incoming Alignment Error (BIAE). This field is used to carry upstream the number of blocks with interleaved bits that have been detected as erroneous (BIE) or the error status for the incoming alignment (BIAE);

– Incoming Alignment Error (IAE);

– bits reserved for future international standardization (RES).

Two bytes support a General Communication Channel (GCC0) between OTU termination points (Figure 6.12). These bytes are located in the columns 11 and 12 of row 1.

6.3.5. *ODU frame over-head structure*

ODU frame is composed of 4 rows and 3,824 columns. Columns 1–14 of row 1 are reserved for FAS, MFAS and the header of the OTU frame (Figure 6.13). The OPU frame is completed by the header ODU OH of the ODU frame located in rows 2–4 and columns 1–14.

PM (Path Monitoring) field, located in columns 10, 11 and 12 of row 3, contains the following subfields (Figure 6.13):

– a Transmitted Trace Identifier (TTI);

– a Bit Interleaved Parity (BIP-8);

– a Backward Defect Indication (BDI);

– a Backward Error Indication (BEI);

– status bits indicating the presence of a maintenance signal (STAT).

APS/PCC (Automatic Protection Switching/Protection Communication Channel) field, located in columns 5–8 in row 4, allows the routing information for the automatic protection switching (Figure 6.13).

FTTL (Fault Type and Fault Location), located in column 14 of row 2 is used to carry a message depicting the fault type and the location, coded on 256 bytes (Figure 6.13).

6 TCM (Tandem Connection Monitoring) fields are defined in columns 5 to 13 of row 2, columns 5 to 13 and in columns 1-9 of row 3 (Figure 6.13).

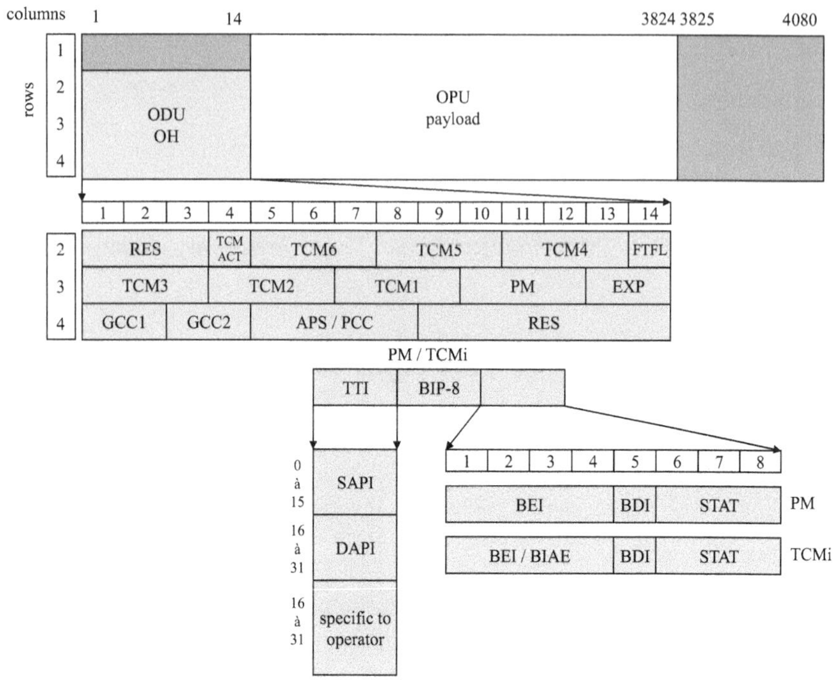

Figure 6.13. *ODU frame overhead structure*

6.3.6. *OPU frame over-head structure*

OPU frame consists of a header and payload corresponding to the Ethernet signal (Figure 6.14):

– columns 15 and 16 are dedicated to the header;

– columns 17 to 3,824 are dedicated to the payload.

The over-head of the OPU frame consists of the following fields (Figure 6.14):

– a Payload Structure Identifier (PSI) including the Payload Type (PT);

– information about the mapping of the Ethernet signal in the payload.

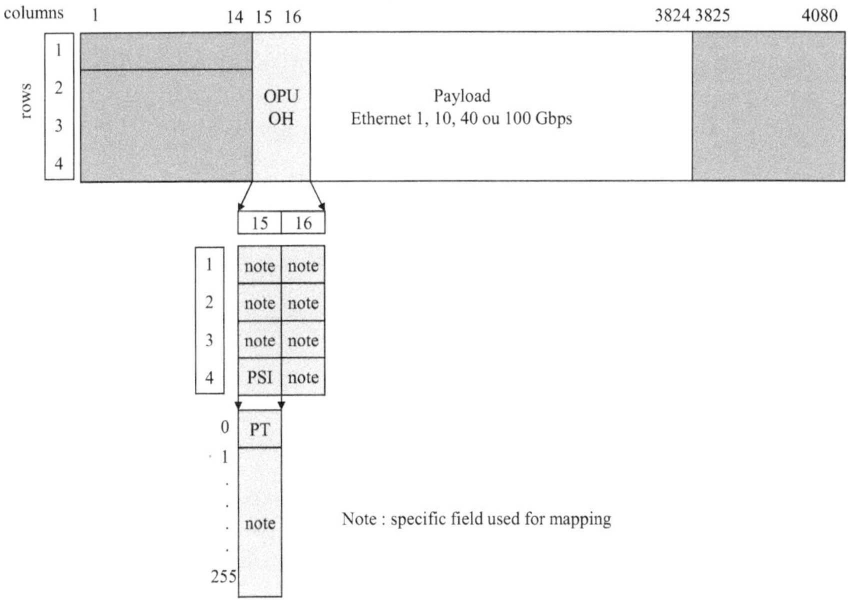

Figure 6.14. *OPU frame overhead structure*

6.4. GFP protocol

6.4.1. *GFP-F protocol*

GFP-F (Generic Frame Procedure - Frame-mapped) adapts the Ethernet frame to the synchronous transmission required for the optical transport network. It also provides a delimitation mechanism of the frame and an error control.

GFP-F frame is composed of the following sets (Figure 6.15):

– a Core Header;

– a Payload Header:

- the payload of the GFP frame consisting of an Ethernet frame,

- optionally, pFCS (payload Frame Check Sequence) field contains an error detection code. This option is not used for the Ethernet frames that have their own error detection code.

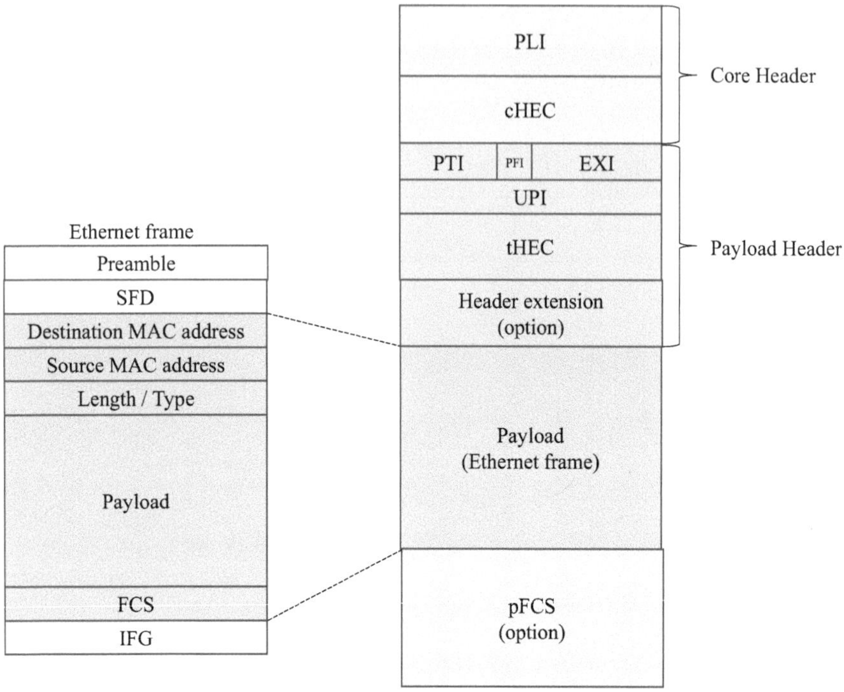

Figure 6.15. *GFP-F frame structure*

The Core Header has two fields:

– PLI (Payload Length Indicator). This field, encoded on 2 bytes, indicates the size of the load constituted by the Payload Header, the Ethernet frame and pFCS field.

– cHEC (core Header Error Check). This field, encoded on 2 bytes, can detect errors in the Core Header.

The Core Header can detect the start of the GFP frame. The receiver calculates the CRC16 of two consecutive bytes and compares it to cHEC field.

If the two values are not identical, the receiver shifts the operation by 1 byte until equality.

When the CRC16 is equal to cHEC field, the receiver recovers the PLI field that determines the end of the GFP frame and the operation is repeated on the next frame.

The Payload Header consists of the following fields:

– PTI (Payload Type Identifier). This field, encoded on 3 bits, indicates if the load contains traffic data or management data;

– PFI (Payload FCS Indicator). This field, encoded on 1 bit, indicates whether pFCS field is present (bit to ONE) or not (bit to ZERO);

– EXI (Extension Header Identifier). This field, encoded on 4 bits, indicates whether an extension of the Payload Header is present or not;

– UPI (User Payload Identifier). This field, encoded on 8 bits, indicates the load type of data (value 0000 0001 for Ethernet);

– tHEC (type Header Error Check). This field, encoded on 2 bytes, can detect errors in the Payload Header.

Ethernet frame encapsulated by the Payload Header does not contain the bytes of the preamble, the Start Frame Delimiter (SDF), allowing the detection of the start of the Ethernet frame, and the Inter frame Gap (IFG).

GFP-F frame is used when the Ethernet frame is mapped into the OPU4, OPU3, OPU2, OPU1 or OPU1 frame.

6.4.2. *GFP-T protocol*

GFP-T (Generic Framing Protocol - Transparent-mapped) reduces the latency associated with storing the Ethernet frame to form the GFP frame. Transparent mode means that the bits are directly encapsulated by the GFP-T header without detecting the presence of the Ethernet frame.

The characters of the 8B/10B code of tributary Ethernet signal to 1 or 10 Gbps are decoded to recover the traffic data and the control data. Note that in the case of GFP-F protocol, control data cannot be transmitted.

The traffic data and the control data are then coded using the 64B/65B code. This will gain an additional rate: 25% overrate in the case of 8B/10B code and 1.56% overrate in the case of 64B/65B code.

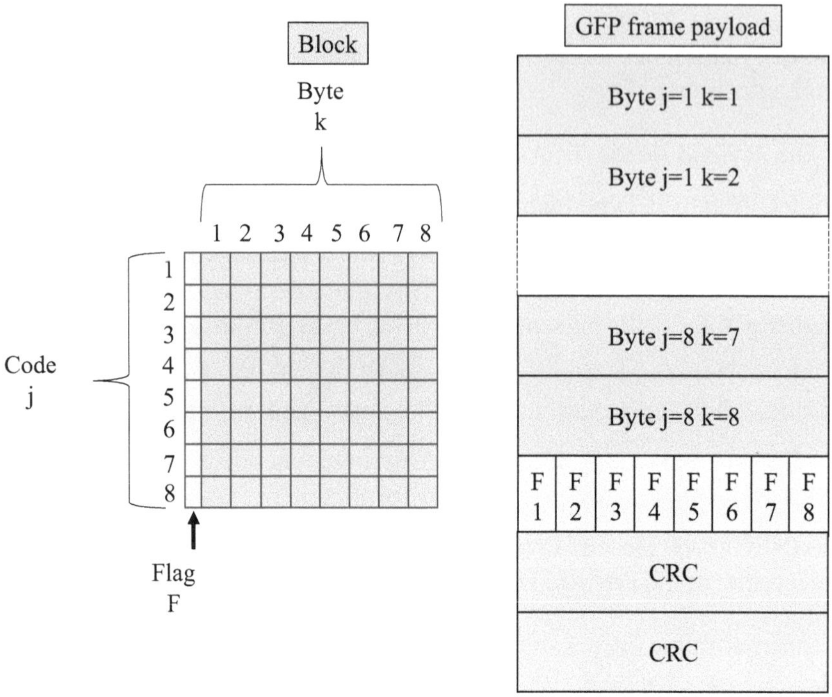

Figure 6.16. *Block structure of the GFP-T frame*

In order to align the bits of the 64B/65B code with the bytes of the OPU frame, a combination of 8 characters of 65 bits, protected by a CRC16, shall enable the constitution of an 11-bytes block (Figure 6.16).

An integer number of blocks is then directly mapped into a GFP-T frame. The number of blocks per GFP frame depends on the rate of the ODU frame. If the rate of the OPU frame increases, the number of blocks per GFP frame decreases.

The header of the GFP-T frame is different from the header of the GFP-F frame by the value (0000 0011) attributed to the UPI field.

As each block is protected by a CRC16, the pFCS field is not used.

7

Quality of Service Principles

7.1. Introduction

Quality of service is the overall effect produced by the performance of a service, which determines the degree of user satisfaction.

Quality of service functions and their associated criteria should be considered from four viewpoints:

– quality of service corresponding to the needs expressed by the user;

– quality of service as offered by the provider: QoS (Quality of Service);

– quality of service actually provided;

– quality of service as perceived by the user: QoE (Quality of Experience).

From the client's point of view, quality of service is expressed using parameters that emphasize the effects felt by the user rather than their origin within the network, wherein the needs may be expressed in a non-technical language.

Quality of service as offered by service providers forms the basis of agreements between the provider and its client, which are defined in a Service Level Agreement (SLA). The quality of service offered by the provider might not correspond to the level required by the client, as considerations such as cost, for example, have to be taken into account.

The quality of service obtained corresponds to a level that is actually reached and provided to the client. It is expressed using measured values and, by comparing them to those stated in the SLA, a level of performance is determined.

The quality of service perceived by the client is expressed in terms of satisfaction level rather than in technical terms. It can be evaluated using surveys among clients or based on comments. The Mean Opinion Score (MOS) is usually defined on a scale 1–5 (Table 7.1).

MOS	Designation
5	Excellent
4	Good
3	Fair
2	Poor
1	Bad

Table 7.1. *User satisfaction level*

7.2. Parameters for the QoS

Network quality parameters can be grouped into three categories, corresponding to the different cases of dysfunction encountered (Table 7.2):

– the client cannot access the network due to the failure of the network;

– the client cannot access the network due to a lack of resources;

– the data characteristics transferred through the network have been degraded.

Categories	Operation quality parameters
Network failure	Availability
Access to resource	Admissibility
Data transfer	Bandwidth, losses, delay, jitter

Table 7.2. *Operation quality parameters*

7.2.1. *Availability*

The availability of a system is the ability of that system to perform a required function at a given point in time. Availability depends on a number of factors:

– the environmental characteristics of the equipment (Table 7.3);

– technical failures of the equipment and the transmission medium;

– data compromise;

– function compromise.

Environmental factors	Examples
Physical disasters	Fire, water damage, landslides, tidal waves, destruction of hardware
Natural events	Climate, seismic, or weather phenomena
Loss of essential utilities	Loss of air conditioning, loss of power
Perturbations due to radiation	Electromagnetic or thermal radiation

Table 7.3. *Environmental factors*

Availability depends on a combination of equipment reliability, transmission medium reliability, maintainability, and maintenance logistics.

The measure of availability (A) is expressed as a percentage of uptime over a given period (Table 7.4).

The measure of unavailability (U) is expressed as a percentage of downtime over a given period:

$$U = 1 - A$$

Availability	Time of unavailability per year
99.999	5 minutes and 11 seconds
99.99	52 minutes and 36 seconds
99.9	8 hours and 46 minutes
99	3 days 15 hours and 40 minutes

Table 7.4. *Availability and unavailability values*

The reliability of a system is the ability of that system to accomplish a required function, under specified conditions of use, for a certain time interval. Measures of the reliability include the failure rate (λ) that, by default, is expressed in default per hour, or the Mean Time Between Failures (MTBF), expressed in hours:

$$MTBF = 1 / \lambda$$

The maintainability of a system is the ability of that system to be maintained in, or restored to, a condition where it can accomplish a required function. The Mean Time to Repair (MTTR) is a measure of maintainability. The major factors that influence maintainability are the detection and the location of faults, and the ease with which a failing element can be reached and replaced.

The maintenance logistics of an organization enable it to provide, on demand, the necessary means in terms of personnel, spare parts, and test equipment to ensure the maintainability of the system.

A key element of network maintainability is the management system that allows for the detection and localization of the fault. This impacts network availability in two ways:

– unavailability of the management system jeopardizes network maintainability if a failure occur therein;

– an illicit action at the management system could cause a service interruption.

The availability (Ai) of an elementary block in a system is governed by the MTBF and MTTR parameters:

$$Ai = MTBF / (MTBF + MTTR)$$

Securing the network is necessary to ensure the availability of the bearer service. This calls upon card redundancy in the equipment, as well as path diversity.

The resultant availability of a system having N blocks in series is equal to the product of elementary uptimes. Thus, arranging blocks in series decreases availability.

Blocks in series: Atotal = A1 × A2 × ... × An

The resultant unavailability of *N* blocks in parallel is equal to the product of elementary downtimes. Arranging blocks in parallel therefore improves availability.

Blocks in parallel: Utotal = U1 × U2 × ... × Un

Network availability is linked to security mechanisms (authentication, confidentiality, integrity), which avoid compromising management and control data.

Compromise of functions results from errors in operation due to non-compliance with instructions or insufficient knowledge of the hardware and Denial of Service, the goal of which is to cause network saturation with the aim of interrupting service.

7.2.2. *Admissibility*

Admissibility is the ability of the network to respond to a resource request from a user. When a user request is accepted, it is offered a resource for its exclusive use.

The implementation of that resource takes place through signaling. The signaling message processor performs Connection Admission Control (CAC) on the user's request.

Two cases may occur when a request is submitted:

– at least one resource allowing the construction of a circuit is available, and the request is immediately fulfilled;

– all the resources are busy with other requests, and the request is rejected.

When all resources are busy, the request is therefore cancelled, and the attempt has to be renewed. Such systems are referred to as loss systems.

Admissibility currently applies mainly to telephone traffic, which corresponds to conversation durations.

Resources are busy for part of the time without necessarily conveying telephone traffic (e.g. time taken by the called party to answer, or non-response from the called party).

The carried traffic Te carried by a circuit is the sum of effective traffic, which corresponds to the conversation time, and ineffective traffic, which corresponds to additional resource occupation.

The offered traffic To, which is made up of all requested calls, whether or not these are fulfilled, may differ from carried traffic Te. Traffic offered is the traffic that would be carried if the number of circuit was sufficient.

The ratio P = (To – Te)/Te represents the proportion of calls which, having found all circuits busy, could not be served. This is the probability of loss, which represents the likelihood of not obtaining a circuit upon request.

Erlang B formula applies to loss systems. This formula provides the relation between traffic offered in Erlang, the number of circuits and the probability of loss of a telephone call.

B formula: $p = (y^n / n!) / \sum_{k=0}^{k=n} (y^k / k!)$

where:

– p = probability of loss;

– y = traffic offered in Erlang;

– n = number of circuit.

Erlang is a unit of measurement of traffic and represents the volume of total traffic in the number of hours during the busy hour.

The Erlang formula applies to telephone service VoLTE (Voice over Long Term Evolution) or videophone service ViLTE (Video over Long Term Evolution) offered by the 4G mobile network, the number of circuits corresponding to the number of bearers allocated.

7.2.3. *Data transfer*

7.2.3.1. *Bandwidth*

Bandwidth is the end-to-end bit rate offered to the user, defined by two parameters:

– Peak Information Rate (PIR), which is the rate measured during an activity period. PIR rate is associated with the size of the data burst PBS (Peak Burst Size) and can be controlled.

– Committed Information Rate (CIR), which is a rate measured over a period including activity and idle periods. CIR rate may be guaranteed and in this case can be controlled. CIR rate is associated with the size of the data burst CBS (Committed Burst Size).

7.2.3.1.1. Audio rate

The audio rate, in the case of VoLTE service, is provided by a codec from the following list:

– Adaptive Multi-Rate (AMR);

– AMR-WB (Wide Band);

– Enhanced Voice Services (EVS).

AMR codec provides the analog / digital conversion of an audio signal in the frequency band 300–3,400 Hz.

AMR codec produces 20-ms frames and the resulting rate can have the following values: 12.2 kbps, 10.2 kbps, 7.95 kbps, 7.40 kbps, 6.70 kbps, 5.90 kbps, 5.15 kbps and 4.75 kbps.

AMR-WB codec has the same technical features as the AMR codec and improves voice quality by increasing the bandwidth (50–7,000 Hz) for the audio signal.

Several configurations (A, B, C) and several rates per configuration are defined:

– configuration A: 6.6 kbps, 8.85 kbps and 12.65 kbps;

– configuration B: 6.6 kbps, 8.85 kbps, 12.65 kbps and 15.85 kbps;

– configuration C: 6.6 kbps, 8.85 kbps, 12.65 kbps and 23.85 kbps.

EVS concept is a new family of codecs whose different types depend on the bandwidth of the audio signal:

– NB (Narrow Band), for an audio signal whose bandwidth is 300–3,400 Hz;

– WB (Wide Band), for an audio signal whose bandwidth is 50–7,000 Hz;

– SWB (Super Wide Band), for an audio signal whose bandwidth is 50–14,000 Hz;

– FB (Full Band), for an audio signal whose bandwidth is 20–20,000 Hz.

The flow rates obtained have the following values: 5.9 kbps, 7.2 kbps, 8 kbps, 9.6 kbps, 13.2 kbps, 16.4 kbps, 24.4 kbps, 32 kbps, 48 kbps, 64 kbps, 96 kbps and 128 kbps.

The rate for the NB type is in the range between 5.9 and 24.4 kbps.

The rate for the WB type is in the range between 5.9 and 128 kbps.

The rate for the SWB type is in the range between 9.6 and 128 kbps.

The rate for the FB type is in the range between 16.4 and 128 kbps.

NB-type and WB-type EVS codecs provide the following improvements compared to the AMR and AMR-WB codecs:

– for the same quality, the rate is decreased;

– for an equivalent rate, the quality is improved.

SWB-type and FB-type EVS codecs are reserved for coding music signals.

7.2.3.1.2. Video rate

The video rate, in the case of VILTE service, is provided by a codec from the following list:

– H.264 codec, with Constrained Baseline Profile (CBP), level 1.2;

– H.265 with Main Profile (MP) level 3.1 Main Tier.

The maximum image size (number of pixels) and the maximum number of frames per second for the CBP Profile level 1.2 can take several values:

– size 176 × 144, 60.6 frame/s;

– size 320 × 240, 20 frame/s;

– size 352 × 288, 15.2 frame/s.

The maximum rate supported by the profile is equal to 384 kbps.

H.265 codec presents the following improvements compared to H.264 codec:

– for the same quality, the rate is decreased;

– for an equivalent rate, the quality is improved.

The maximum image size (number of pixels) and the maximum number of frames per second for the MP profile level 3.1 Main Tier can take several values:

– size 720 × 576, 75 frame/s;

– size 960 × 540, 60 frame/s;

– size 1,280 × 720, 33.7 frame/s.

7.2.3.1.3. Data rate

The maximum rate for data using TCP (Transmission Control Protocol) is limited by the size of the anticipation window RWIN and the Round Trip Time between sending TCP segment and receiving acknowledgment.

$$\text{Rate} \leq \text{RWIN/RTT}$$

The value of the anticipation window RWIN is less than or equal to 65,565 bytes for standard operation of TCP. Optionally, the scaling can increase this value.

TCP protocol implements the mechanisms “slow start” and “congestion avoidance” to discover the rate that the network can flow.

The “slow start” mechanism is to use a transmission window which doubles the value for each received acknowledgment, until the avoidance threshold of congestion.

The “congestion avoidance” mechanism is to evolve the transmission window linearly when the avoidance threshold of congestion is reached.

In case of detection loss of segment loss, “fast recovery” mechanism used to return the segment when the source receives duplicated acknowledgments for the same segment.

During a segment loss, to avoid restarting the “slow start” mechanism and reducing sharply the rate, “fast retransmission” mechanism is to maintain the transmission window after the receipt of the transmitted segments.

The algorithms present in the operating systems (CUBIC for Linux and CCTP for Windows) can maintain a rate as close as possible to the available bandwidth.

7.2.3.2. *Loss ratio*

The loss ratio is the ratio of the number of blocks correctly received to the number of blocks emitted. The loss ratio is mainly related to the bit error rate and to the congestion of a node in the network.

Congestion corresponds to the saturation of the queues at the node, which occurs when the demand for traffic (incoming traffic to the node) exceeds available resources (output ports rate).

The network may also introduce defects in traffic restitution:

– Out-of-order packet delivery. This situation can arise when several paths are possible between the source and destination, with each path having a different transfer time.

– Erroneous insertion. This situation can arise when an error in a header is not detected.

The bit error rate is the ratio of binary elements correctly received and the total number of received binary elements. This measure is used to qualify the performance of the transmission function between two devices.

The bit error rate is related to:

– the transmission link budget, which corresponds to the difference between transmitter power and receiver sensitivity. The receiver sensitivity is provided for a given error rate;

– the perturbations affecting the transmission channel (channel deformation, interference, signal fading).

7.2.3.3. *Delay*

Delay has a direct impact upon user satisfaction, depending on the application. It is composed of delays at the hosts and in the network. The user's sensitivity to delay arises in various ways:

– time required to establish a particular service from the initial user request;

– time required to receive the data once the service has been established.

Network-induced delay can be split into a fixed delay and a variable delay. The fixed delay depends on the following:

– data packetization time, which depends on the size of the encapsulated data;

– data serialization time, which depends on the output interface rate;

– propagation time, which depends on the distance between the transmitter and receiver. The average propagation time is 3 μs/km (free-space propagation) or 5 μs/km (propagation in an optical fiber or metallic pair);

– processing time, which depends on the functions to be performed in order to make the connection at the network node.

The variation in delay is primarily caused by the variation of data storage time in queues. The value used is the average value of the delay.

7.2.3.4. *Jitter*

Variable delay within the network causes jitter. For applications with very low tolerance to delay variations (real-time applications, such as voice or video), buffering measures on the received data make it possible to eliminate delay variation at the cost of introducing an additional fixed delay.

Real-time Transport Protocol (RTP) contains a Time Stamp (TS) field, which allows us to replay the data at the same rate as that of the source.

On the other hand, the storage time in the buffer is limited, and any packet is deleted, if the value of jitter is superior to the storage time.

7.2.3.5. *Service classes*

Service classes correspond to a combination of limit values of performance for a set of streams having comparable characteristics.

ITU (International Telecommunication Union) Recommendation Y.1541 defines the values reported in Table 7.5 for networks implementing Internet Protocol (IP).

There are two types of real-time services: the interactive service (service class 0), corresponding to the telephone or videophone, and non-interactive service (service class 1), corresponding to a voice mail service, unidirectional television transmitted in unicast flow (Video On Demand) or in multicast flow (Live).

These two types of service have the same constraints regarding the loss and the jitter. By cons, they differ in terms of the delay.

		Service classes					
		0	1	2	3	4	5
Parameters	Loss	10^{-3}	10^{-3}	10^{-3}	10^{-3}	10^{-3}	Not specified
	Delay	100 ms	400 ms	100 ms	400 ms	1 s	Not specified
	Jitter	50 ms	50 ms	Not specified	Not specified	Not specified	Not specified

Table 7.5. *Service classes*

There are thousands of data applications, each with its own limitations in terms of rate, packet loss and delay.

Some applications use the TCP transport protocol whose acknowledgment mechanism can retransmit lost packets.

Others use the UDP transport protocol that does not offer this feature. In this case, the acknowledgment mechanism is carried over the application level.

Also, some applications are delay sensitive, others not. The nature of the flow is also variable: uniform traffic or traffic burst, significant or low bandwidth. On the other hand, thc variation of the delay is not noticeable by the user.

All these applications can be grouped into four classes of service:

– interactive data and control data belong to the service class 2;

– transactional data belong to the service class 3;

– bulk data belong to the service class 4;

– background data belong to the service class 5.

Interactive data relate to protocols such as TELNET and instant messaging. Prescription is severe with regard to the delay that must be less than 250 ms. For example, to TELNET, the return of the echo character must be virtually instantaneous. Interactive applications typically have a volume to transfer relatively small.

Control data include data necessary for the establishment of the user plane (e.g. routing protocols, telephony signaling protocols). The requirement is particularly strict regarding the delay of the data. For example, the delay of the call signaling data directly impacts the time between sending the phone dialing and receiving the ring back tone.

Transactional data concern the database consultations. The main prescription here is to give the user a feeling of imminence, showing that the transaction is proceeding. In this case, a delay of several seconds at most is desirable. Transactional applications can have a volume to transfer relatively large.

Bulk data include file or database transfers and their replication. From the moment that there is an indication that the transfer is in progress, it makes sense to tolerate a period slightly longer than for transactional data.

Background data are for Internet applications such as browsing to access a webpage or e-mail. Class-5 applications to which these belong do not specify guarantee on performance parameters. It corresponds to the Best Effort service.

7.3. QoE characterization

7.3.1. *Application to voice*

QoE quality on the telephone service is represented by the voice quality perceived by the listener that depends on the degradation of the sound signal from the mouth of the talker and the ear of the listener.

E-Model, developed by the ITU (International Telecommunication Union), aims to evaluate the mouth-to-ear quality in a conversation, as perceived by the receive-side user both as a listener and as a talker. The raw result is the evaluation factor R, which combines all the transmission parameters that apply to the data transfer.

The evaluation factor R can take values from 0 to 100, where $R = 0$ represents an extremely poor quality and $R = 100$ an extremely good quality (Table 7.6).

The estimated MOS for the conversational quality on a scale of 1–5 can be obtained from the factor R by means of the following formula:

$$MOS = 1 + 0.035 \times R + R(R - 60)(100 - R) \times 7 \times 10^{-6}$$

R value	MOS	User satisfaction
90	4.34	Very satisfied
80	4.03	Satisfied
70	3.60	A few unsatisfied users
60	3.10	Many unsatisfied users
50	2.58	Almost all unsatisfied

Table 7.6. *Correspondence between the evaluation factor R and the MOS*

This evaluation factor R is composed of the following terms:

$$R = Ro - Is - Id - Id - Ie + A$$

Ro represents the basic signal-to-noise ratio and includes the effect of noise sources such as circuit noise and room noise. The basic signal-to-noise ratio Ro is defined by the following relation:

$$Ro = 15 - 1.5 \times (SLR + No)$$

SLR (Send Loudness Rating) corresponds to the send loudness as measured at the telephone set microphone.

No term is the power addition of the different noise sources:

- Nc is the sum of all circuit noise powers;
- Nos is the circuit noise caused by room noise on the send side;
- Nor is the circuit noise caused by room noise on the receive side;
- Nfo represents the background noise at the receiver.

Is represents all the impairments that occur more or less simultaneously with the voice signal, including, for example, a speech level (telephone volume) that is too high or too low, a non-optimum local effect or quantization noise.

It is subdivided into three specific impairment factors:

$$Is = Iolr + Ist + Iq$$

where:

– Iolr factor represents the decrease in quality caused by too-low values or too-high of the overall loudness rating. The overall loudness rating represents the attenuation in the transmission chain;

– Ist factor represents the impairment caused by a local effect not reaching an optimum value;

– Iq factor represents the impairment caused by quantization distortion.

The local effect is the acoustic coupling of sounds that are being transmitted from the microphone to the earpiece of the same handset. The coupling is brought about by the mechanical effects of the telephone handset, facial bones, the air around the talker's head and the connection of the subscriber to the equipment.

The impairment corresponding to quantization distortion is inherent to the PCM (Pulse Code Modulation) codec. This impairment is perceived as noise accompanying the received voice signal (that is it is not perceived in an idle channel).

Id represents the impairments caused by delay and echo effects. It is also subdivided into three factors:

$$Id = Idte + Idle + Idd$$

where:

– Idte factor provides an estimate for the impairments due to echo for the talker;

– Idle factor represents impairments due to echo for the listener;

– Idd factor represents the impairment caused by excessive absolute delay.

Electrical echo occurs when the communication links together analog telephone sets, which typically involves two-wire to four-wire converters. Electrical echo is non-existent when the communication links together digital telephone sets.

The impairment due to electrical echo for the talker depends on two factors: the delay and the level of the voice signal reflected toward the talker. Perceived quality diminishes when the delay and/or the level of the received echo signal increase.

The use of devices, such as echo cancellers, enables a reduction in the level of echo received and contributes to improving perceived quality. The installation of such devices is recommended on all connections whose one-way delay along the talker echo path is greater than 25 ms, which is systematically the case in mobile networks.

As the delay increases, it starts to have an influence on conversational interactivity. The delay affecting the reply of a remote correspondent becomes perceptible from 150 ms upward. It should be noted that the telephone codec and the RTP buffer can introduce a non-negligible delay.

The equivalent equipment impairment factor, Ie, represents the impairments caused by low-rate codecs. It also represents impairment due to random-distribution packet losses.

Packet Loss Concealment (PLC) is a technique that enables the effect of a packet loss to be masked. The Sequence Number field of the RTP header which encapsulates the voice, allows detection, at the receiving end, a lost block that will be synthesized by the codec.

The advantage factor A compensates for impairment factors when there is some benefit in providing access to the user, for example mobility or connections in hard to reach regions.

E model was adapted to take into account the improvement of the intrinsic quality of wideband codecs (50–7,000 Hz). The factor of transmission R evaluation factor can be set to 0 –129.

Calculating the MOS parameter remains unchanged with a maximum value equal to 4.5 for R = 100. For values of R > 100, the MOS value is maintained at 4.5.

7.3.2. *Application to video*

QoE quality on the video telephony service is represented by three parameters:

– $S_q(V_q)$ parameter indicates the level of voice quality, influenced by the quality of the video signal. The determination of this parameter is currently under study.

– $V_q(S_q)$ parameter indicates the level of video quality influenced by the quality of the audio signal. The determination of this parameter is currently under study.

– MM_q parameter, calculated from MM_{SV} and MM_T parameters:

$$MM_q = m_1 \times MM_{SV} + m_2 \times MM_T + m_3 \times MM_{SV} \times MM_T + m_4$$

MM_{SV} is the index of the quality of audio and video signals.

MM_T is the index for the end-to-end delay of the audio and video signals.

The value of the parameters m_1–m_4 is shown in Table 7.7.

MM_{SV} index is calculated from the quality of the audio signal (S_q) and the video signal (V_q):

$$MM_{SV} = m_5 \times S_q + m_6 \times V_q + m_7 \times S_q \times V_q + m_8$$

The value of the parameters m_5–m_8 is shown in Table 7.7.

Coefficients	Screen size 4.2 inch	Screen size 2.1 inch
m_1	-4.457×10^{-1}	-6.966×10^{-1}
m_2	6.638×10^{-1}	-8.127×10^{-1}
m_3	4.042×10^{-1}	4.562×10^{-1}
m_4	2.321	3.003
m_5	-3.255×10^{-1}	-1.638×10^{-1}
m_6	3.309×10^{-1}	3.626×10^{-1}
m_7	1.494×10^{-1}	1.494×10^{-1}
m_8	5.457×10^{-1}	5.456×10^{-1}
m_9	-3.235×10^{-4}	-1.251×10^{-4}
m_{10}	3.915	3.763
m_{11}	-1.377×10^{-3}	-1.065×10^{-3}
m_{12}	0.000	1.465×10^{-2}
m_{13}	-1.095×10^{-3}	-1.002×10^{-3}
m_{14}	0.000	0.000

Table 7.7. *Parameters m_1–m_{14}*

S_q quality of the audio signal takes into account the following parameters:

– the end-to-end delay;

– the type and the rate of the audio codec;

– the packet loss rate;

– the level of the echo for the talker.

V_q quality of the video signal takes into account the following parameters:

– the cnd-to-end delay;

– the characteristics of the video codec (type, size, bit rate, frame rate, screen size);

– the packet loss rate.

MM_T index is calculated from the AD parameter relating to the absolute delay of audio and video signals and the MS parameter relating to the synchronization of audio and video signals.

$$MM_T = \max \{AD+MS, 1\}$$

where:

– $AD = m_9 (T_S + T_V) + m_{10}$;

– $MS = \min \{m_{11} (T_S - T_V) + m_{12}, 0\}$ if $T_S \geq T_V$;

– $MS = \min \{m_{13} (T_V - T_S) + m_{14}, 0\}$ if $T_S < T_V$;

– T_S = absolute delay of the audio signal;

– T_V = absolute delay of the video signal.

The value of the parameters m_9–m_{14} is shown in Table 7.7.

8

Quality of Service Mobile Network

8.1. Flow and bearer

A Service Data Flow (SDF) is a group of flows (IP packets) used as part of a service, while the EPS (Evolved Packet System) bearer refers to all the SDF flows belonging to the same class of service (Figure 8.1).

The Traffic Flow Template (TFT) allows the User Equipment (UE) for outgoing packets and the PDN Gateway (PGW) for incoming packets to make the correspondence between, on one hand, the IP flows, and, on the other hand the EPS bearer (Figure 8.1).

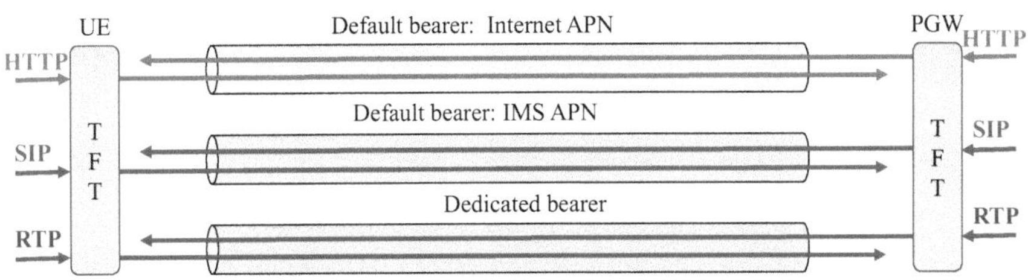

Figure 8.1. *Flow characterization. For a color version of this figure, see www.iste.co.uk/perez/4gmobile.zip*

8.2. QoS characteristics

The SDF flow is associated with a QoS (Quality of Service) with characteristics provided by the Policy and Charging Control (PCC).

The SDF flow can have a Guaranteed Bit Rate (GBR) or a non-GBR.

The EPS bearer inherits QoS level of the SDF flow it carries.

Table 8.1 provides the QoS characteristics of the EPS bearer, related to the two bearer families.

QoS characteristics	GBR	Non-GBR
QCI (QoS Class Identifier)	✓	✓
ARP (Allocation and Retention Priority)	✓	✓
GBR (Guaranteed Bit Rate)	✓	
MBR (Maximum Bit Rate)	✓	
APN-AMBR (Aggregate Maximum Bit Rate)		✓
UE-AMBR		✓

Table 8.1. *QoS characteristics*

8.2.1. *QCI parameter*

QoS Class Identifier (QCI) defines a service class and specifies the characteristics of QoS (priority level, delay and packet loss rate) associated with the service class (Table 8.2).

QCI from 1 to 4 are assigned to GBR bearers.

QCI from 5 to 9 are assigned to non-GBR bearers.

The scheduling of traffic data carried out at the level of the eNB (evolved Node Base station) SGW (Serving Gateway) and PGW entities is based on the QCI priority level.

The flow with the most priority (Priority = 1) is the one that carries the telephone signaling SIP (Session Initiation Protocol) used for the IMS (IP Multimedia Subsystem) network.

The bit rate control is conducted for each bearer at the eNB and PGW entities for incoming data in the EPS network.

The bit rate control is done from the GBR and MBR parameters for guaranteed bit rate bearers.

QCI	Resource type	Priority	Delay (ms)	Packet loss rate	Examples of services
1	GBR	2	100	10^{-2}	Voice
2		4	150	10^{-3}	Video Calling
3		3	50	10^{-3}	Games
4		5	300	10^{-6}	Video
5	Non-GBR	1	100	10^{-6}	SIP Signaling
6		6	300	10^{-6}	Video, Internet
7		7	100	10^{-3}	Voice, Video, Games
8		8	300	10^{-6}	Video, Internet
9		9			

Table 8.2. *QCI parameters*

The bit rate control is done from the APN-AMBR and UE-AMBR parameters for non-GBR type bearers. This control is performed for aggregated bit rates of non-GBR bearers of a mobile.

The APN-AMBR parameter controlled by the PGW entity corresponds to the maximum bit rate authorized for all SDF flows of a mobile using non-GBR bearers, at the PGW entity level.

The UE-AMBR parameter controlled by the eNB entity corresponds to the maximum authorized bit rate for all SDF flows of a mobile using non-GBR bearers, at the eNB entity level.

8.2.2. *ARP parameter*

The pre-emption implemented at the eNB and PGW entity level corresponds to the ARP (Allocation and Retention Priority) parameter that defines the following information:

– Priority: this parameter determines the priority level associated with pre-emption. This priority level is independent of that set for the QCI parameter.

– Pre-emption Capability: this parameter is used for the establishment of a new session, if the resource is not available. This parameter determines whether or not a new session can pre-empt an existing session.

– Pre-emption Vulnerability: this parameter is used by the existing session. This parameter is compared to the Pre-emption Capability parameter of the new session to determine whether the existing session can be pre-empted or not.

8.2.3. *Applying QoS*

Figure 8.2 illustrates the application of quality of service to SDF flow and EPS bearer.

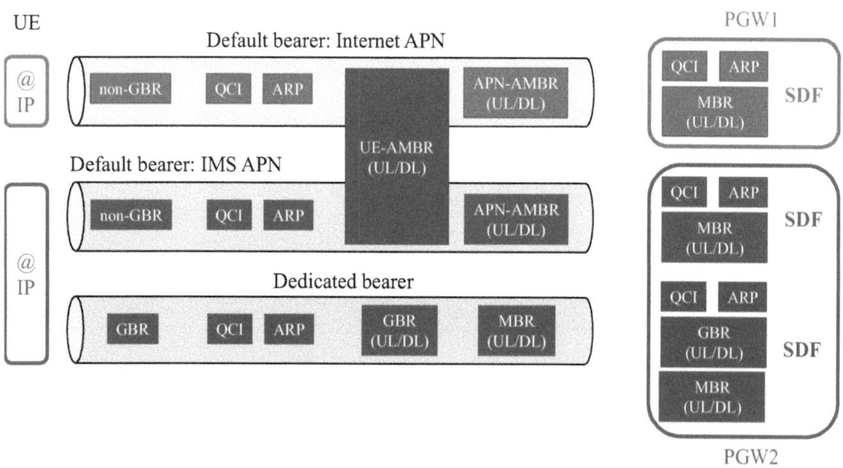

Figure 8.2. *QoS parameters of the SDF flow and the EPS bearer. For a color version of this figure, see www.iste.co.uk/perez/4gmobile.zip*

The mobile has two IP addresses, one is assigned by the PGW1 entity for Internet service, the other by the PGW2 entity for telephone service.

The mobile is connected to the entity PGW1 by a default bearer in which SDF pass several SDF flows which have the same QoS characteristics.

The mobile is connected to the PGW2 entity by a default bearer for the transport of telephone signaling and a dedicated bearer for the voice transmission.

The QoS parameters (QCI, ARP, APN-AMBR) relating to default bearers are stored in the Home Subscriber Serve (HSS). These values can be changed by the Policy and Charging Rules Function (PCRF).

The QoS parameters (QCI, ARP, GBR, MBR) relating to dedicated bearers are stored in the Subscription Profile Repository (SPR) entity associated with the PCRF entity.

The Mobility Management Entity (MME) replaces the UE-AMBR parameter provided by the HSS entity by the sum of the different APN-AMBR parameters, provided it is less than the value indicated by the HSS entity.

8.3. Configuring QoS

8.3.1. *Default-bearer establishment*

The procedure for the default-bearer is described in section 3.3.1. Figure 8.3 extends the description of this process by introducing the configuration of QoS.

1) The procedure for default-bearer establishment is triggered by the mobile when it sends to the MME entity the ESM PDN CONNECTIVITY REQUEST message.

At the end of the authentication phases, the security of NAS (Non-Access Stratum) protocol and the update of the location, the MME entity retrieves the Access Point Name (APN) and the QoS characteristics.

Internet APN: QCI = 9, ARP = 7, APN-AMBR (UL/DL) = unlimited/unlimited

IMS APN: QCI = 5, ARP = 6, APN-AMBR (UL/DL) = 100 kbps/100 kbps

The MME entity selects the SGW entity from its group and the PGW entity from a Domain Name Server (DNS) resolution of the APN.

2) The MME entity sends the GTPv2-C CREATE SESSION REQUEST message to create a context at the SGW entity.

The GTPv2-C CREATE SESSION REQUEST message contains the EPS bearer identifier, the International Mobile Subscriber Identity (IMSI), the APN and the QoS characteristics.

3) The SGW entity sends the GTPv2-C CREATE SESSION REQUEST message to create a context at the PGW entity.

The GTPv2-C CREATE SESSION REQUEST message contains the EPS bearer identifier, the IMSI, the APN and the QoS characteristics.

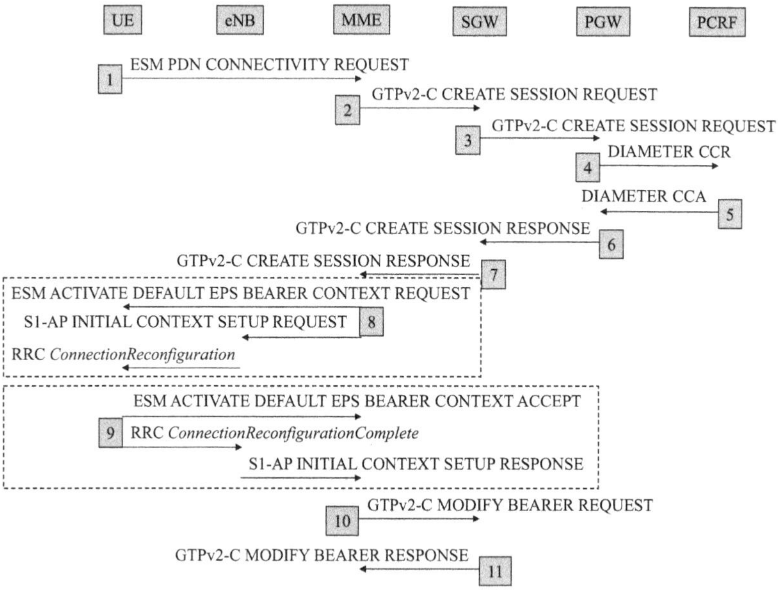

Figure 8.3. *Default-bearer establishment*

4) The PGW entity sends to the Policy and Charging Rules Function (PCRF), the DIAMETER CCR (Credit-Control-Request) message containing the IMSI, the APN and the QoS characteristics.

The PCRF entity compares the profile of the mobile with the rules defined for the network and stored in the Subscription Profile Repository (SPR).

5) The PCRF entity responds to the PGW entity with the DIAMETER CCA (Credit-Control-Answer) message that contains the QoS characteristics and the SDF table defining the incoming flow (DL) and the outgoing flow (UL).

> Internet SDF flow: QCI = 9, ARP = 5, MBR (UL/DL) = unlimited/unlimited, SDF table (UL/DL) = (UE IP, *, *, *, *)/(*, UE IP, *, *, *)
>
> IMS SDF flow: QCI = 5, ARP = 6, MBR (UL/DL) = 100 kbps/100 kbps, SDF table (UL/DL) = (UE IP, *, SIP, *, UDP)/(*, UE IP, SIP, *, UDP)

PGW1 and PGW2 entities develop the characteristics of EPS bearer from the characteristics of the SDF flow.

> Internet EPS bearer (PGW1): QCI = 9, ARP = 5, MBR (UL/DL) = unlimited/unlimited, TFT table (UL/DL) = (UE IP, *, *, *, *)/(*, UE IP, *, *, *)
>
> IMS EPS bearer (PGW2): QCI = 5, ARP = 6, APN-AMBR (UL/DL) = 100 kbps/100 kbps, TFT table (UL/DL) = (UE IP, *, SIP, *, UDP)/(*, UE IP, SIP, *, UDP)

6) The PGW entity responds to the SGW entity with the GTPv2-C CREATE SESSION RESPONSE message that contains the EPS bearer identifier, the QoS characteristics and the TFT table for outgoing flow (UL) only.

> Internet EPS bearer: QCI = 9, ARP = 5, MBR (UL) = unlimited, TFT table (UL) = (UE IP, *, *, *, *)
>
> SIP EPS bearer: QCI = 5, ARP = 6, APN-AMBR (UL) = 100 kbps, TFT table (UL) = (UE IP, *, SIP, *, UDP)

7) The SGW entity responds to the MME entity with the GTPv2-C CREATE SESSION RESPONSE message that contains the EPS bearer identifier, the QoS characteristics and the TFT table for outgoing flow (UL) only.

8) The MME entity responds to the mobile with the ESM ACTIVATE DEFAULT EPS BEARER CONTEXT REQUEST message that contains the EPS bearer identifier, APN, QoS characteristics and TFT table for outgoing flow (UL) only.

The ESM ACTIVATE DEFAULT EPS BEARER CONTEXT REQUEST message is carried by the S1-AP INITIAL CONTEXT SETUP REQUEST message on the S1-MME interface and by the RRC *ConnectionReconfiguration* message on the LTEU-u interface.

The S1-AP INITIAL CONTEXT SETUP REQUEST message allows the creation of the mobile context at the eNB entity level and contains the EPS bearer identifier, the QoS characteristics and the value of the UE-AMBR (unlimited).

9) The mobile confirms receipt of the ESM ACTIVATE DEFAULT EPS BEARER CONTEXT REQUEST message by sending the ESM ACTIVATE DEFAULT EPS BEARER CONTEXT ACCEPT message.

The ESM ACTIVATE DEFAULT EPS BEARER CONTEXT ACCEPT message is carried by the RRC *ConnectionReconfigurationComplete* message on the LTE-Uu interface and by the S1-AP INITIAL CONTEXT SETUP RESPONSE message on the S1-MME interface.

The RRC *ConnectionReconfigurationComplete* message acknowledges the reception of the RRC *ConnectionReconfiguration* message.

10) The MME entity updates the S1 bearer by the GTPv2-C MODIFY BEARER REQUEST messages sent to the SGW entity.

11) The SGW entity acknowledges the message received with the GTPv2-C MODIFY BEARER RESPONSE message.

8.3.2. *Dedicated-bearer establishment*

The procedure for dedicated-bearer establishment is described in section 3.3.3. Figure 8.4 extends the description of the process by introducing the configuration of QoS.

The dedicated bearer is coupled to the default bearer assigned to the telephone signaling, in that the terminations of the bearers are the same for both types of bearer.

1) The establishment of dedicated bearer supporting voice is triggered by the Application Function (AF) when it sends the PCRF entity the DIAMETER AAR (Authenticate-Authorize-Request) message containing the flow characteristics to establish (codec, bit rate, identifier of the IP flow).

2) The PCRF entity consults its SPR database to retrieve the traffic profile to apply and transmits the DIAMETER RAR (Re-Auth-Request) message to the PGW entity containing the QoS characteristics and the SDF table defining the incoming flow (DL) and the outgoing flow (UL).

SDF flow: QCI = 1, ARP = 7, GBR (UL/DL) = 88 kbps/88 kbps, MBR (UL/DL) = 88 kbps/88 kbps, SDF table (UL/DL) = (UE IP, *, RTP, *, UDP)/(*, UE IP, RTP, *, UDP)

The PGW entity establishes the characteristics of the EPS bearer from the characteristics of the SDF flow.

EPS bearer: QCI = 1, ARP = 7, GBR (UL/DL) = 88 kbps/88 kbps, MBR (UL/DL) = 88 kbps/88 kbps, TFT table (UL/DL) = (UE IP, *, RTP, *, UDP)/(*, UE IP, RTP, *, UDP)

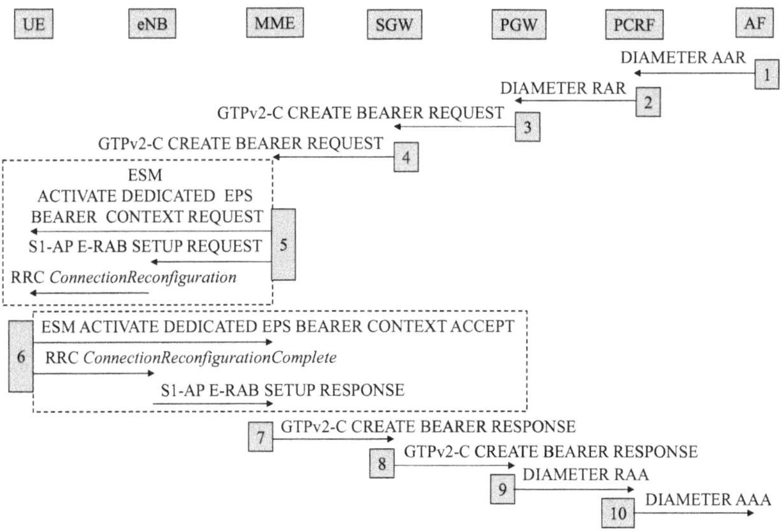

Figure 8.4. *Dedicated bearer establishment*

3) The PGW entity sends to the SGW entity the GTPv2-C CREATE BEARER REQUEST message containing the identifier of the default bearer allocated to the telephone signaling and the identifier of the dedicated bearer allocated to the voice, the QoS characteristics for the incoming flow (DL) and the outgoing flow (UL), and the TFT table for the outgoing flow (UL) only.

EPS bearer: QCI = 1, ARP = 7, GBR (UL/DL) = 88 kbps/88 kbps, MBR (UL/DL) = 88 kbps/88 kbps, table TFT (UL) = (UE IP, *, RTP, *, UDP)

4) The SGW entity sends to the MME entity the GTPv2-C CREATE BEARER REQUEST message containing the identifier of the default bearer allocated to the telephone signaling and the identifier of the dedicated bearer allocated to the voice, the QoS characteristics for the incoming flow (DL) and the outgoing flow (UL) and the TFT table for the outgoing flow (UL) only.

5) The MME entity sends to the mobile the ESM ACTIVATE DEDICATED EPS BEARER CONTEXT REQUEST message carried by the S1-AP E-RAB SETUP REQUEST message on the S1-MME interface and by the RRC *ConnectionReconfiguration* message on the LTE-Uu interface.

The ESM ACTIVATE DEDICATED EPS BEARER CONTEXT REQUEST message contains the identifier of the default bearer allocated to the telephone signaling and the identifier of the dedicated bearer allocated to the voice, the QoS characteristics and the TFT table for the outgoing flow (UL) only.

PS bearer: QCI = 1, ARP = 7, GBR (UL) = 88 kbps, MBR (UL) = 88 kbps, TFT table (UL) = (UE IP, *, RTP, *, UDP)

The S1-AP E-RAB SETUP REQUEST message contains the identifier of the default bearer allocated to the telephone signaling and the identifier of the dedicated bearer allocated to the voice, the QoS characteristics for the incoming flow (DL) and the outgoing flow (UL).

EPS bearer: QCI = 1, ARP = 7, GBR (UL/DL) = 88 kbps/88 kbps, MBR (UL/DL) = 88 kbps/88 kbps

6) The mobile responds to the MME entity by the ESM ACTIVATE DEDICATED EPS BEARER CONTEXT ACCEPT message carried by the RRC *ConnectionReconfigurationComplete* message on the LTE-Uu

interface and by the S1-AP E-RAB SETUP RESPONSE message on the S1-MME interface.

7) The MME entity responds to the SGW entity with the GTPv2-C CREATE BEARER RESPONSE message.

8) The SGW entity responds to the PGW entity with the C-GTPv2 CREATE BEARER RESPONSE message.

9) The PGW entity responds to the PCRF entity with the DIAMETER RAA (Re-Auth-Answer) message.

10) The PCRF entity responds to the P-CSCF entity with the DIAMETER AAA (Authenticate-Authorize-Answer) message.

9

Quality of Service Transport Network

9.1. IP technology

The DiffServ (Differentiated Services) model provides differentiated treatment for each class of service consisting of an aggregate flow. Traffic classes comprise a set of flows identified by the following parameters: source and destination IP (Internet Protocol) addresses, source and destination port numbers and PROTOCOL field of the IP header.

This bundling by class of service limits the number of behaviors to be managed by the router.

The DiffServ model thus defines, at each router, the mechanism to be applied to a set of flows belonging to a class of service for the allocation of resources (buffers, rate) and for the Per-Hop Behavior (PHB) including congestion avoidance and queue scheduling.

9.1.1. *DSCP field*

The first specification of the IPv4 header reserved the ToS (Type of Service) field to carry out priority management. The first three bits specify the precedence level. The next three bits provide an indication for minimizing Delay, maximizing Throughput, or maximizing Reliability (Figure 9.1).

The DSCP (DiffServ Code Point) field uses the six bits in the IPv4 or IPv6 header to define a new marking of the packet. Routers use the DSCP marking to select a PHB corresponding to a particular treatment for the data transfer (Figure 9.1).

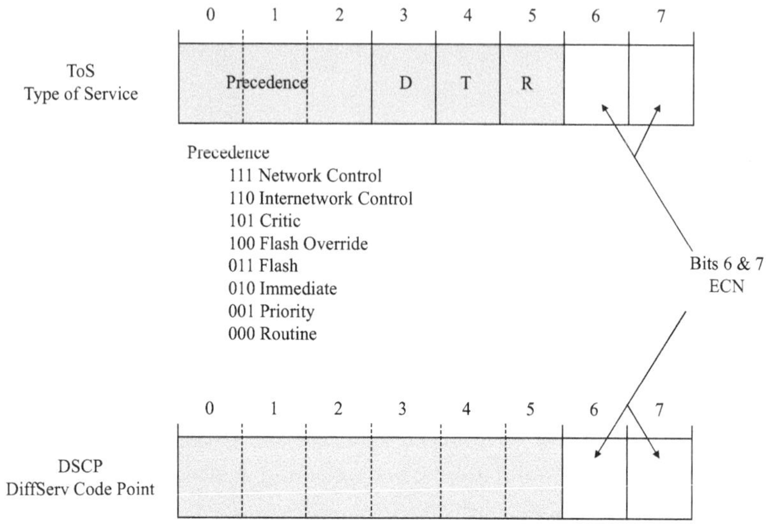

Figure 9.1. *ToS and DSCP fields of the IP header*

The value of the DSCP mark is defined at the ingress of the DiffServ domain, that is at the input port of the ingress router. Flow classification is the operation that enables to gather flows for which a particular marking will be applied.

It should be noted that the value of the DSCP mark may be defined upstream from the DiffServ domain. In that case, the ingress router only carries out a rate policing based on the value of the DSCP field.

The value of the DSCP mark is used at the output ports of the DiffServ domain routers to assign the flow to queues. The congestion avoidance and the scheduling mechanism are associated with the PHB.

Before the advent of DiffServ, the ToS field was used to differentiate flows. The CS (Class Selector) marking maintains upward compatibility with the ToS field with regard to the PRECEDENCE bits.

The CS marking does not provide compatibility with the other three bits in the ToS field (delay, throughput, and reliability) (Table 9.1).

The values of the DSCP mark are in the form XXX000, where the XXX bits correspond to the value of the PRECEDENCE bits. The DSCP mark defines eight classes. An order relation is established between the classes: for any two classes CSx and CSy, if x>y, the packets belonging to CSx should be delivered with a higher probability than the one allocated to the packets belonging to CSy.

PHB	CS0	CS1	CS2	CS3	CS4	CS5	CS6	CS7
DSCP	000 000	001 000	010 000	011 000	100 000	101 000	110 000	111 000

Table 9.1. *DSCP field: CSx mark*

The EF (Expedited Forwarding) class is used to obtain a low loss ratio, a low delay and low jitter (Table 9.2). These characteristics are obtained if aggregate traffic is not stored (or not much of it is stored) in queues. This arrangement is only possible if the rate of such aggregate traffic is lower than that allocated at the output port.

PHB	EF
DSCP	101 110

Table 9.2. *DSCP field: EF mark*

The AF (Assured Forwarding) class specifies four service classes. The first three bits of the DiffServ Code Point field encode the value of the traffic class. The remaining three bits encode three precedence levels for each traffic class. A packet that belongs to class i, with a precedence level j, is marked with the DSCP value corresponding to the AFij marking (Table 9.3).

PHB	AF11	AF12	AF13	AF21	AF22	AF23
DSCP	001 010	001 100	001 110	010 010	010 100	010 110
PHB	AF31	AF32	AF33	AF41	AF42	AF43
DSCP	011 010	011 100	011 110	100 010	100 100	100 110

Table 9.3. *DSCP field: AFx mark*

Within a given traffic class, packets of precedence x are delivered with a higher probability than the one given to packets of precedence y, if $x < y$.

Packets that belong to one traffic class should be transferred independently from packets belonging to another traffic class. The router must allocate an amount of resources (queue size, rate) to each traffic class.

Table 9.4 indicates the DSCP marking that can be used for different applications.

DSCP mark	Types of applications
CS6	Control data (routing protocols (OSPF, BGP), DNS, DHCP protocols, SNMP alarm messages)
EF	Bidirectional voice (telephony)
CS5	Control data (telephone signaling (SIP))
AF4x	Bidirectional video (video telephony, videoconference)
CS4	Interactive data
AF3x	Unidirectional video and voice (on-demand video)
CS3	Unidirectional video and voice (broadcast video)
AF2x	Transactional data
CS2	Control data (management protocols (SNMP))
AF1x	Bulk data (FTP)
CS0	Background data (HTTP, SMTP, POP3)

Table 9.4. *DSCP marking for applications*

Table 9.5 shows the mechanisms for PHB to be applied depending on the value of the DSCP mark.

DSCP mark	Policing	Congestion avoidance	Scheduling
CS6	Not applicable	WRED	WRR, WFQ
EF	Token bucket srTCM	Tail Drop	PQ
CS5	Token bucket srTCM	Tail Drop	WRR, WFQ
AF4x	Token bucket trTCM	WRED	WRR, WFQ
CS4	Token bucket srTCM	WRED	WRR, WFQ
AF3x	Token bucket trTCM	WRED	WRR, WFQ
CS3	Token bucket srTCM	WRED	WRR, WFQ
AF2x	Token bucket trTCM	WRED	WRR, WFQ
CS2	Token bucket srTCM	WRED	WRR, WFQ
AF1x	Token bucket trTCM	WRED	WRR, WFQ
CS0	Not applicable	WRED	WRR, WFQ

Table 9.5. *PHB mechanisms associated with the DSCP mark*

The mechanisms (Token bucket srTCM and trTCM) for rate control (policing) are described in section 9.4.

The mechanisms for avoidance congestion (WRED) and scheduling queues (PQ, WRR, WFQ) are described in section 9.5.

9.1.2. *DiffServ architecture*

A DiffServ domain comprises border routers and internal routers. A DiffServ domain defines a flow marking policy and its associated PHB.

A DiffServ region is made up of several DiffServ domains. Each DiffServ domain can implement different policies. The border routers connect the DiffServ domain to other domains.

The internal routers only connect to other internal or border DiffServ routers within the same DiffServ domain (Figure 9.2).

The border routers act either as ingress or egress nodes, depending on the traffic direction. Traffic enters a domain via an ingress router and leaves a domain via an egress router.

An ingress router makes sure incoming traffic into the DiffServ domain conforms to the Service Level Agreement (SLA).

An egress router carries out conditioning functions on the traffic transmitted to the following user, in accordance with the terms of the SLA.

The ingress router first processes the flow from the input port (Figure 9.3). The MF CLASSIFIER function splits that flow into microflows based on the allocation to different traffic classes.

The various entities of 4G mobile network perform marking the DSCP field, carried out upstream of the DiffServ domain, which prevents the realization of the MF CLASSIFIER function at the ingress router.

This mark is a mapping between the values of the QCI (QoS Class Identifier) parameter and the DSCP field (Table 9.6).

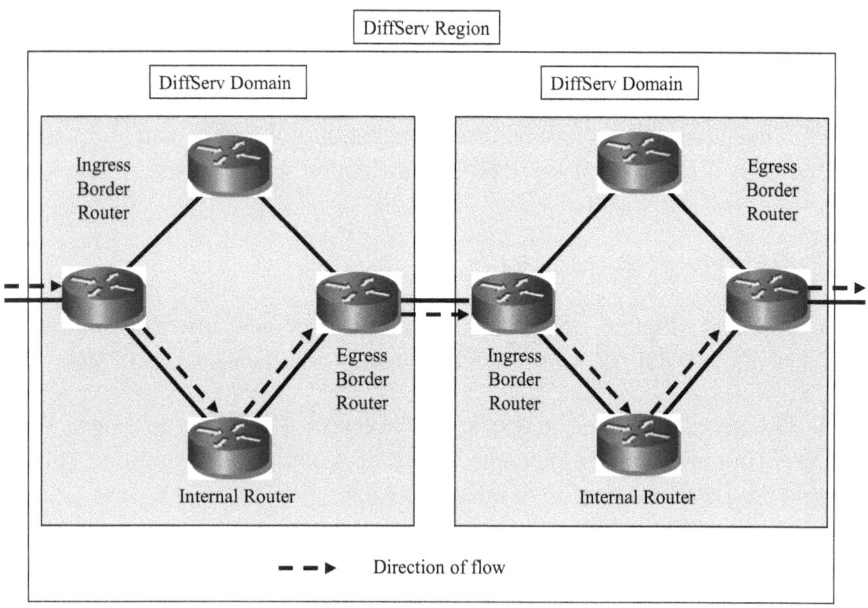

Figure 9.2. *DiffServ architecture: domain and region*

QCI	DSCP
1	EF
2	
3	
4	AF41
5	AF31
6	AF32
7	AF21
8	AF11
9	CS0

Table 9.6. *Mapping between DSCP and QoS parameters*

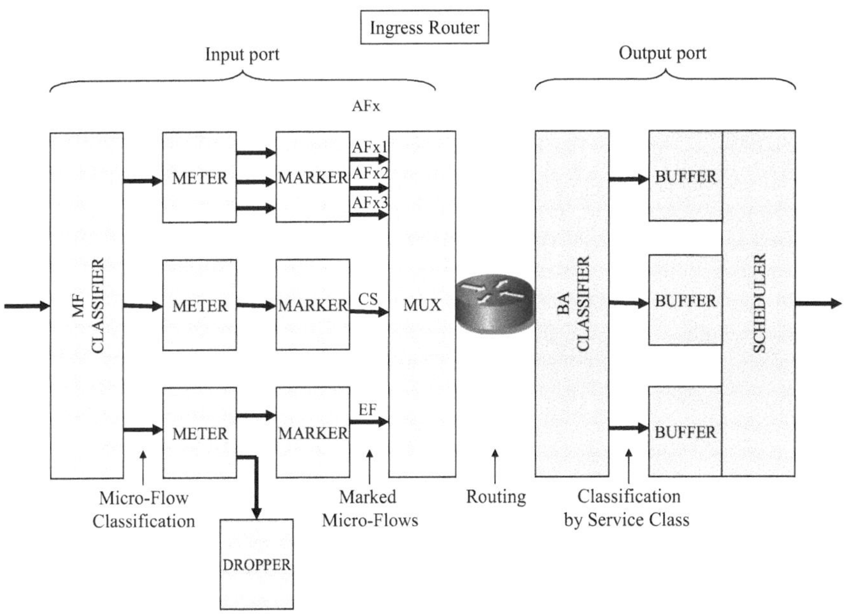

Figure 9.3. *DiffServ architecture: ingress border router*

For each microflow, the METER function checks conformity with the SLA by using the token bucket mechanism. A difference in treatment occurs between the EF traffic and the AF traffic. Non-conforming packets from the EF family are dropped (DROPPER); on the other hand, non-conforming packets from AF have their loss probability increased according to the precedence level (AF 11, AF 12 and AF 13).

When all the microflows have been marked, the multiplexing (MUX) function reconstructs the input flow so that it can be transmitted to the routing function.

At the second stage, the ingress router performs processing to construct the flow at the output port (Figure 9.3). The BA (Behavior Aggregate) CLASSIFIER function sorts the flow based on the value of the DSCP field and assigns each packet to the corresponding queue.

The SCHEDULER function carries out scheduling among queues to construct the output flow.

The internal routers in the DiffServ domain perform packet routing and classification (BA CLASSIFIER) and queue scheduling (SCHEDULER) at the output ports (Figure 9.4).

The egress router performs the same functions as an internal router. It may additionally assign a new DSCP marking to accommodate a new DiffServ domain (Figure 9.4).

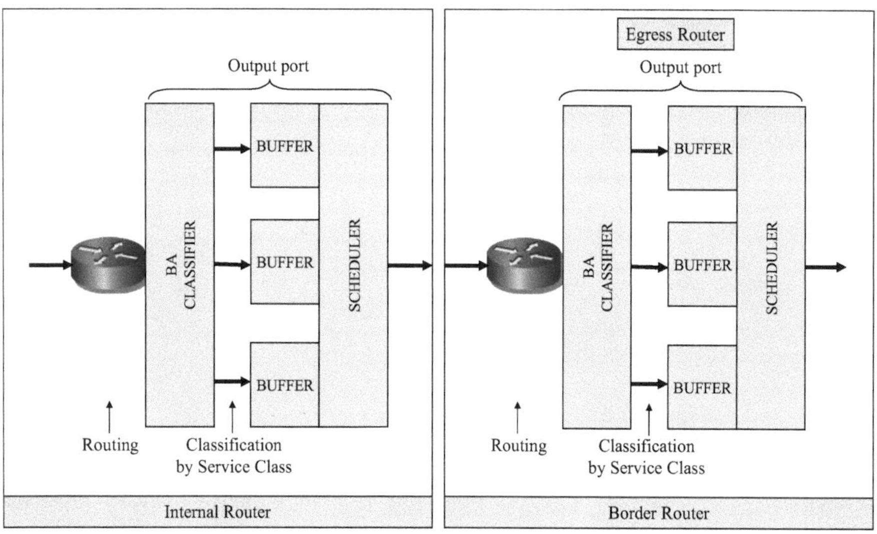

Figure 9.4. *DiffServ architecture: internal and egress border routers*

9.2. Ethernet technology

The Ethernet protocol defined a 3-bit field to encode the service class to which the flow belongs. As is the case with DiffServ, these 3 bits are used to address, at the switch level, the behavior mechanisms associated with the classes of service.

These 3 bits are carried by an additional 4-byte field, which is used to carry out VLAN (Virtual Local Area Network) marking (Figure 9.5):

– two bytes insert the Tag Protocol Identification field so as to signal that the VLAN marking has been added (value 81-00 in hex notation);

– two bytes insert the Tag Control Identification (TCI) field.

The TCI field contains the following information (Figure 9.5):

– the PCP (Priority Code Point) field encoded on 3 bits. It provides the mark allocated to the frame;

– the CFI (Canonical Format Indicator) field encoded on 1 bit. A switch will always set this value to zero;

– the VID (VLAN Identification) field encoded on 12 bits. It identifies the VLAN to which the source of the frame belongs. The VID field is also called Q-VLAN. Q-VLAN tagging allows to partitioning the flow within the Ethernet network.

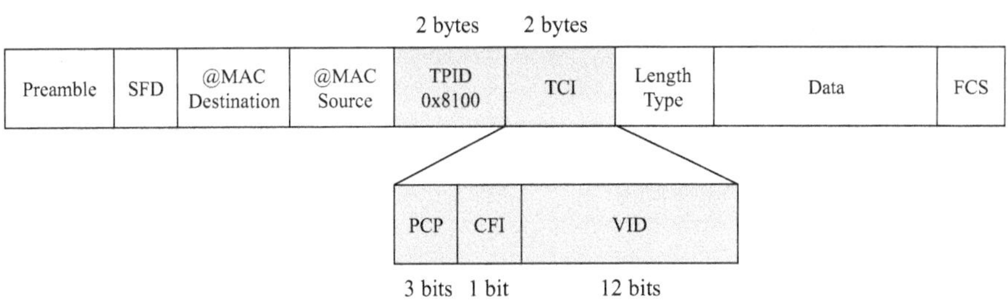

Figure 9.5. *Marking of the Ethernet header*

Table 9.7 indicates the PCP marking used for the different classes of service.

Priority level	Acronym	Class of service
1	BK	Background
2	–	–
0	BE	Best effort
3	EE	Excellent effort
4	CL	Controlled load
5	VI	Video
6	VO	Voice
7	NC	Network control

Table 9.7. *Marking the classes of service*

It is possible to have a correspondence between, on the one hand, the value of the DSCP field in the IP header, and on the other hand, the value of the PCP field in the Ethernet header (Table 9.8).

DSCP value (decimal)	Classes DSCP	PCP value (decimal)
0	CS0 (000000)	0
8	CS1 (001000)	1
10	AF11 (001010)	1
16	CS2 (010000)	2
18	AF21 (010010)	2
24	CS3 (011000)	3
26	AF31 (011010)	3
32	CS4 (100000)	4
34	AF41 (100010)	4
40	CS5 (101000)	5
46	EF (101110)	5
48	CS6 (110000)	6
56	CS7 (111000)	7

Table 9.8. *Correspondence between DSCP and PCP marks*

Ethernet frame, whether marked as Q-VLAN or not, can receive a new S-VLAN (Service VLAN) mark. The Q-VLAN marking changes denomination and becomes C-VLAN (Customer VLAN).

The S-VLAN mark enables the operator to provide flow partition mechanisms on a per-client basis and preserve, within its network, the C-VLAN marking carried out by thc client.

The S-VLAN marking is identical to the Q-VLAN marking, with one exception: the CFI bit is replaced by the DEI (Drop Eligible Indicator) bit (Figure 9.6). This bit defines a precedence level. If the bit has a value of ONE, it has a lower precedence level. If the bit has a value of ZERO, it has a higher precedence level.

The priority level encoded in the PCP field addresses the qucue scheduling mode. The precedence level encoded in the DEI field is used for the congestion avoidance mechanism (e.g. the WRED mechanism).

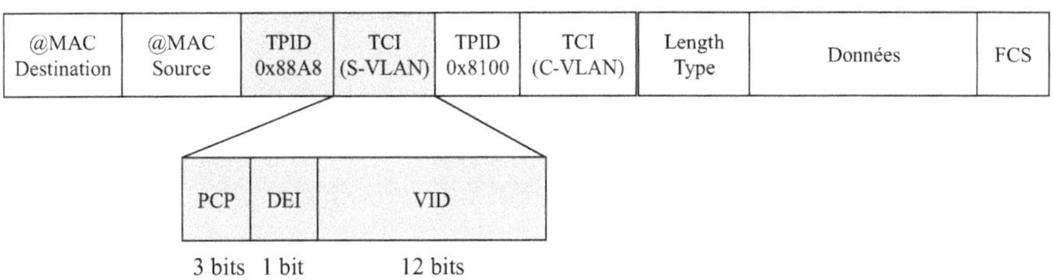

Figure 9.6. *Double marking of the Ethernet header*

It is also possible for the priority level and precedence level to be encoded only in the PCP field. The eight values of the PCP field can be used in different ways (Table 9.9):

– eight priority levels and no precedence level (8P0D);

– seven priority levels and, for one priority level, and one precedence level (7P1D);

– six priority levels and, for two priority levels, and one precedence level (6P2D);

– five priority levels and, for three priority levels, and one precedence level (5P3D).

Priority		7		6		5		4		3		2		1		0	
Precedence		0	1	0	1	0	1	0	1	0	1	0	1	0	1	0	1
PCP	8P0D	7		6		5		4		3		2		1		0	
	7P1D	7		6		5	4	5	4	3		2		1		0	
	6P2D	7		6		5	4	5	4	3	2	3	2	1		0	
	5P3D	7		6		5	4	5	4	3	2	3	2	1	0	1	0

Table 9.9. *Encoding of the PCP field*

9.3. MPLS technology

9.3.1. *Support for DiffServ*

In an MPLS network supporting DiffServ, the LSR (Label Switch Routers) entities run the same traffic processing mechanisms as those defined within the DiffServ architecture.

The MPLS network introduces two types of virtual circuits LSP (Label Switching Path):

– a virtual circuit of E-LSP (EXP-inferred-class LSP) type carries several classes of service simultaneously;

– a virtual circuit of L-LSP (Label-inferred-class LSP) type carries a single class of service.

For the E-LSP model, the LSR entities use the EXP field in the MPLS header to determine behavior. The EXP field contains three bits and can therefore assume eight possible values.

Consequently, a virtual circuit of E-LSP type can only encode eight classes of service. The value of the EXP field is defined on entering the MPLS network, for example, based on a correspondence with the DSCP (DiffServ Code Point) field of the IP header.

For the L-LSP model, the LSR entities use the label and EXP field of the labeled packet to determine the PHB.

It is thus possible to have a one-to-one correspondence between, on one hand, the value of the DSCP field in the IP header, and, on the other hand, the value of the label representing the class of service, the EXP field representing the precedence level for the AF classes.

Thirteen labels are required to encode all values of the DSCP field (Table 9.10).

DSCP field	Class of service	Label Class of service	EXP field precedence
EF	EF	L1	000
AF43	AF4	L2	011
AF42			010
AF41			001
AF33	AF3	L3	011
AF32			010
AF31			001
AF23	AF2	L4	011
AF22			010
AF21			001
AF13	AF1	L5	011
AF12			010
AF11			001
CS7	CS7	L6	000
CS6	CS6	L7	000
CS5	CS5	L8	000
CS4	CS4	L9	000
CS3	CS3	L10	000
CS2	CS2	L11	000
CS1	CS1	L12	000
CS0	CS0	L13	000

Table 9.10. *Correspondence between marks of IP and MPLS headers*

9.3.2. *Interaction between markings*

Three types of interaction between the PHB behavior of the DiffServ domain, for IP and MPLS networks, have been defined: pipe model, short pipe model and uniform model.

In the pipe model, the LSR entities apply the PHB based on the marking according to the type of LSP, E-LSP or L-LSP. The LSR entities therefore ignore the DSCP marking in the IP header in order to define the PHB (Figure 9.7).

The MPLS network cannot implement the PHP (Penultimate Hop Popping) function, as the egress LSR entity no longer has the marking for determining the PHB.

The short pipe model, like the pipe model, makes it possible to interconnect different DiffServ domains and to ensure transparence on the DSCP field of the IP header. The difference lies in the determination of the PHB of the egress LSR.

The PHB is obtained not from the marking of the LSP, but from the DSCP marking of the IP header (Figure 9.8). This arrangement therefore enables the LSP to use the PHP function (Figure 9.9).

The uniform model is applied when the DiffServ domain is extended to the MPLS network (Figure 9.10). In this case, the packet has the same behavior inside and outside the MPLS network. This arrangement, therefore, enables the LSP to use the PHP function (Figure 9.11).

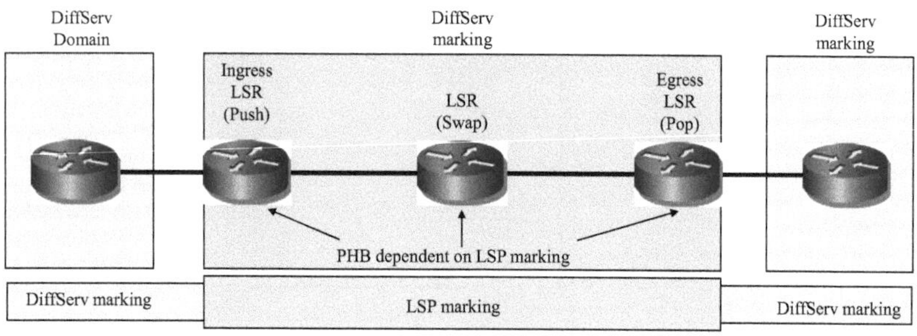

Figure 9.7. *Pipe model*

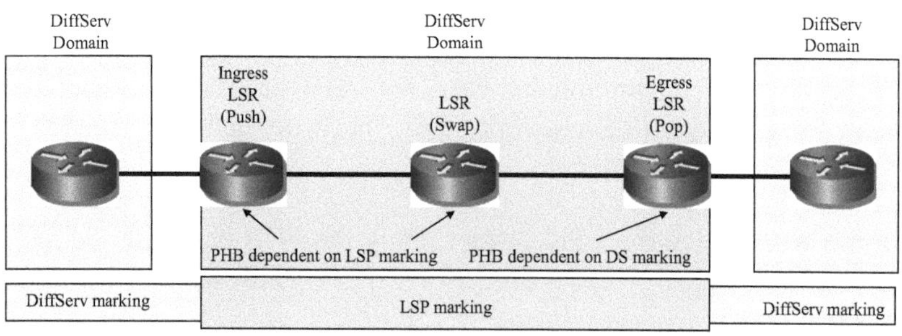

Figure 9.8. *Short pipe model without PHP function*

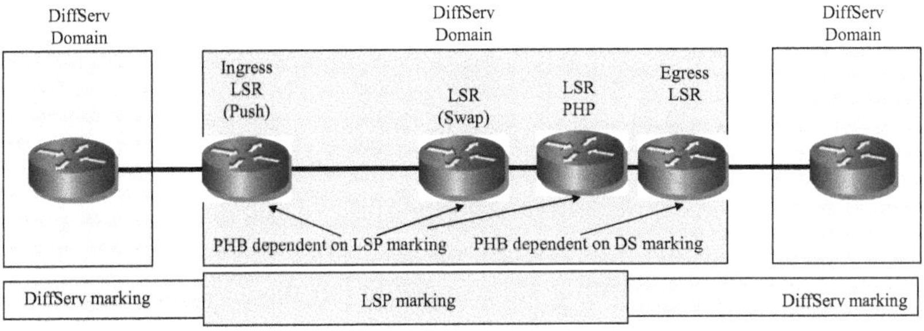

Figure 9.9. *Short pipe model with PHP function*

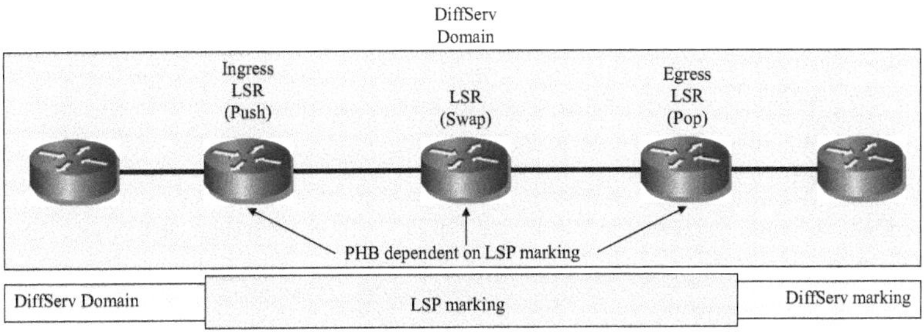

Figure 9.10. *Uniform model without PHP function*

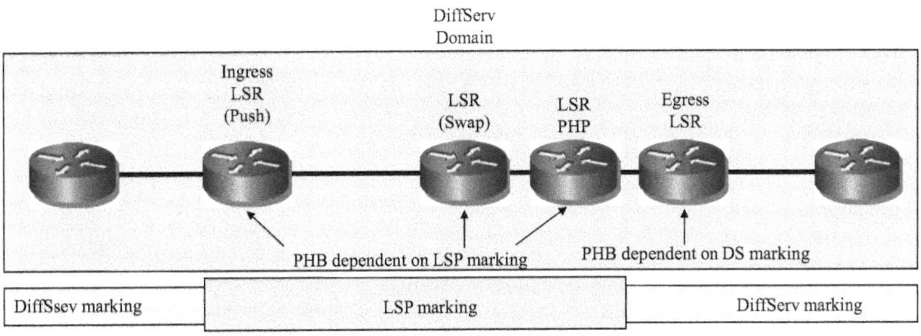

Figure 9.11. *Uniform model with PHP function*

9.4. Rate control

Traffic policing aims to solve the problem of overall congestion in the network. This function allows the traffic at the network ingress to be limited by discarding or downgrading packets that fail to meet specifications of the SLA established between the client and the provider.

Traffic shaping allows a reduction in the size of packet bursts by spreading them in time. The first reason that drives shaping traffic is the presence of the policing function at the ingress of the network. The second reason is linked to the congestion of queues at the output port of the equipment. Spreading traffic enables a reduction in the occupancy of these queues.

Token bucket characterizes a flow using two parameters `r` and `b` (Figure 9.12). The first term represents the Committed Information Rate (CIR); the second one indicates the Maximum Burst Size (MBS) the source can generate. Using this mechanism, a flow is characterized by its average rate and the acceptable temporary deviation from that rate.

For the token bucket mechanism, `r` represents the rate at which the bucket fills up and `b` denotes its size. At an initial point in time, the bucket is filled with tokens. Upon the arrival of a packet of length L, L tokens are removed from the bucket.

Between the arrivals of two packets, the level of the bucket rises at a rate of `r` tokens per second. The filling stops when the level of the bucket reaches `b`. As long as there are tokens in the bucket, it is considered that the flow conforms to specifications.

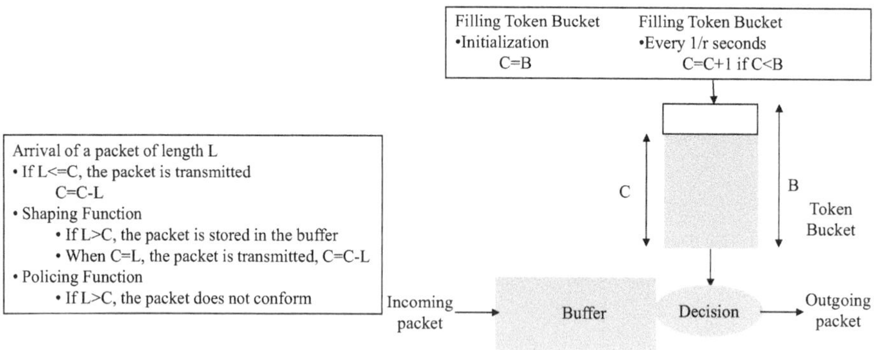

Figure 9.12. *Token bucket*

The token bucket is used to perform the policing function on ingress. It enables the detection of packets that do not conform to specifications. When it is detected that a packet does not conform, it can be either discarded or marked with a lesser precedence level.

An additional memory stores packets that do not conform to specifications. It is used to perform the traffic shaping function.

The srTCM (single rate Three Color Marker) characterization meters the IP packet flow and marks it with three colors: GREEN, YELLOW or RED (Figure 9.13). The color results in a value for the DSCP field in the IP header.

The marking is based on the Committed Information Rate (CIR) and on two burst sizes: the Committed Burst Size (CBS) and the Excess Burst Size (EBS).

The packet is colored GREEN if the burst size does not exceed the CBS parameter, YELLOW if it exceeds the CBS parameter but not the EBS parameter and RED otherwise, if it exceeds the EBS parameter.

The srTCM characterization is used when policing is performed only over the burst length and not on the peak rate.

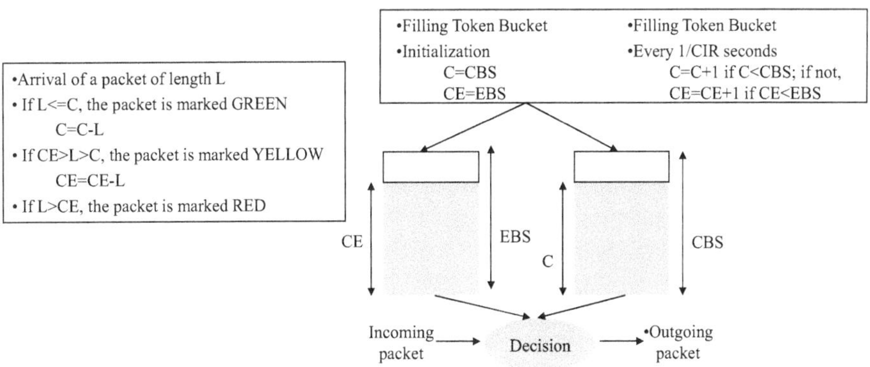

Figure 9.13. *srTCM token bucket*

The trTCM (two rate Three Color Marker) characterization meters the IP packet flow and marks it with three colors: GREEN, YELLOW or RED (Figure 9.14). The packet is marked RED if it exceeds the Peak Information

Rate (PIR). Otherwise, the packet is marked either YELLOW or GREEN, depending on whether or not the burst size exceeds the CBS parameter.

The trTCM characterization is used when the peak rate PIR needs to be controlled irrespective of the average committed rate (CIR).

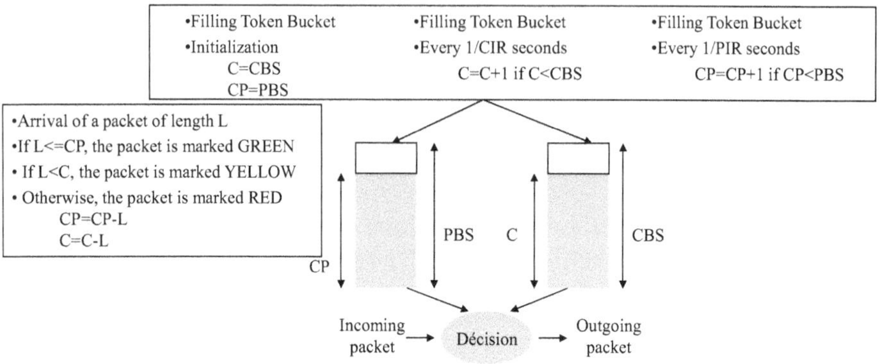

Figure 9.14. *trTCM token bucket*

9.5. Behavior

DSCP mark in IP header, PCP in Ethernet header or EXP in MPLS header is used to address an output port queue for which a PHB behavior is defined, including avoiding congestion and scheduling queues.

9.5.1. *Congestion avoidance*

Traditionally, the Tail Drop discarding function is used in network nodes. If the incoming rate is greater than the outgoing rate, packets accumulate in the queue until it reaches its maximum size. From then on, all incoming packets are dropped.

While this method is the simplest one to implement, it has a major drawback: if a large number of packets are dropped consecutively, this causes the synchronization of several TCP flows simultaneously entering the Slow Start mode. This results in oscillations in the network load, which is linked to a variation of occupancy in queues.

The RED (Random Early Discard) drop function is based on the average queue size (avg). This is calculated using the instantaneous occupancy (wq) of the queue and the weight (q) assigned to each new measurement. The calculation takes into account the periods of queue inactivity, assuming that (m) packets of minimum size could have been transmitted.

```
Calculation of average occupancy avg
Initialization
avg = 0
At the arrival of each packet, if queue not empty,
avg = (1 - wq)avg + wqq
Otherwise, m = time for which the queue has been empty
avg = (1 - wq)mavg
```

RED uses an increasing random function to decide whether a packet should be dropped. Random drop only occurs when the average queue size lies between two thresholds, thmin and thmax.

```
Calculation of drop probability
if thmin <= avg <= thmax, count = count + 1
pb = pmax(avg - thmin) / (thmax - thmin)
pa = pb / (1 - count * pb)
with probability pa:
       drop the packet
       count = 0
if thmax <= avg
drop the packet
count = 0
```

The variable (count) counts packets accepted into the queue since the last drop. The variable (pa) represents the probability of dropping the packet. It increases linearly from 0 to pmax as the average varies from thmin to thmax. Beyond thmax, all packets are dropped.

The RED algorithm thus prevents TCP sources from simultaneously initiating the Slow Start mechanism.

The WRED (Weighted Random Early Discard) drop function relies on the use of different RED parameters for different flows (Figure 9.15).

By assigning different values to pmax, thmin and thmax, it is possible to offer a differentiated drop distribution.

The calculation of average occupancy remains the same as in RED mechanism. On the arrival of each packet, the avg parameter is updated irrespective of the type of flow. On the other hand, differentiation takes place in the calculation of the drop probability (pa).

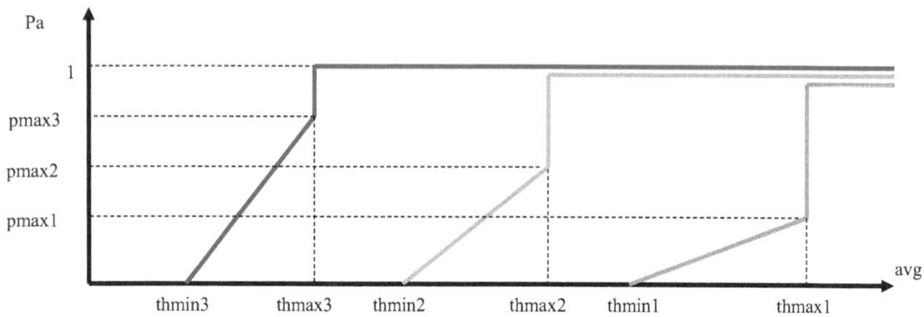

Figure 9.15. *WRED mechanism*

9.5.2. *Queue scheduling*

In its simplest concept, scheduling is about reading a single queue using the FIFO (First-In First-Out) algorithm. This algorithm does not provide any mechanism for fair bandwidth allocation or for establishing a priority level for several flows or aggregate flows.

RR (Round Robin) scheduling consists of cycling sequentially through the queues. If a queue is empty, the scheduler moves on to the following queue. If a queue contains at least one packet, this is transmitted over the output port, irrespective of its length.

The fairness afforded by this mechanism depends on the average packet size in each queue. A queue containing packets that are twice as large as the others will obtain twice as much bandwidth.

DRR (Deficit Round Robin) scheduling makes it possible to ensure the fairness of each queue with regard to allocated bandwidth (Figure 9.16).

A token (D) is allocated to each queue when the scheduler passes it.

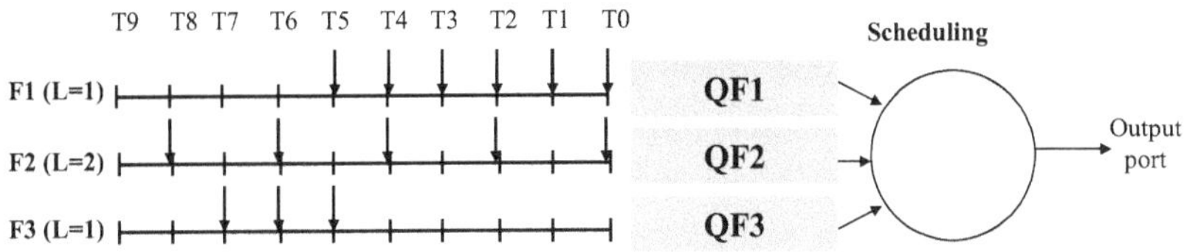

T	D[1]	D[2]	D[3]	Q(F1)	Q(F2)	Q(F3)	Scheduled
0	0+1−1	0	0	P10	P20	–	P10
1	0+1−1	0+1	0	P11	P20	–	P11
2	0	1+1−2	0	P12	P22	–	P20
3	0	0	0	P12, P13	P22	–	P20(cont)
4	0+1−1	0	0	P12, P13, P14	P22,P24	–	P12
5	0	0+1	0+1−1	P13, P14,P15	P22,P24	P35	P35
6	0+1−1	0	0	P13, P14,P15	P22,P24,P26	P36	P13
7	0	1+1−2	0	P14, P15	P22,P24,P26	P36,P37	P22
8	0	0	0	P14, P15	P24,P26	P36,P37	P22(cont)
9	0	0	0+1−1	P14, P15	P24,P26	P36,P37	P36

Figure 9.16. *DRR scheduling*

If the size of the packet at the beginning of the queue is greater than the number of accumulated tokens, the packet is not transmitted and the tokens are available for use at the next round.

If the size of the packet at the beginning of the queue is less than or equal to the number of accumulated tokens, the packet is transmitted via the output port, and the number of accumulated tokens is decremented by a value corresponding to the size of the transmitted packet.

WRR (Weighted Round Robin) scheduling introduces the notion of weight on queues to provide a differentiation of bandwidth per queue. The weighting is implemented if different token values are assigned to each queue.

WFQ (Weighted Fair Queuing) scheduling relies on the notion of timestamps that are calculated using the weight assigned to the flow and the packet length (Figure 9.17).

The timestamp of the flow in queue QF1 varies as follows:

– the value of the timestamp of the first packet is equal to 3. This value corresponds to the ratio of the packet length (L) (value 1) to the weight (r = 1/3) assigned to the flow;

– the value of the timestamp of the second packet is equal to 6. This value corresponds to the value of the timestamp of the first packet (value equal to 3) incremented by a value (equal to 3) corresponding to the ratio of the length (L) (value 1) of the second packet to the weight ($r = 1/3$);

– the value of the following packets' timestamps is incremented by applying the same process.

The packet transmitted over the output port corresponds to the one with the lowest timestamp value.

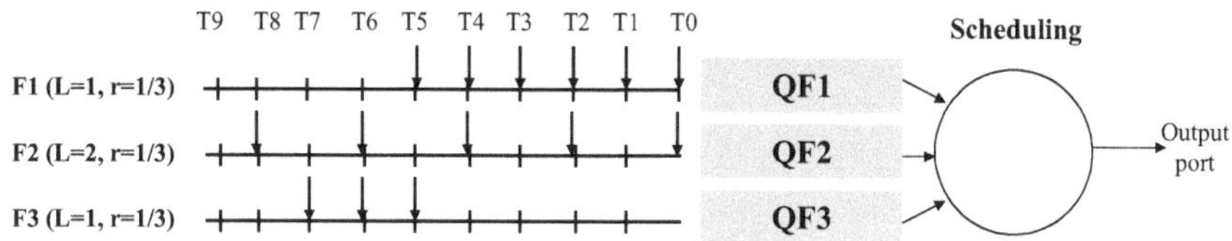

T	E[1]	E[2]	E[3]	Q(F1)	Q(F2)	Q(F3)	Scheduled
0	max(0,0)+3	max(0,0)+6	0	P10 (3)	P20 (6)	–	P10
1	max(1,3)+3	0	0	P11 (6)	P20 (6)	–	P20
2	max(2,6)+3	max(2,3)+6	0	P11 (6), P12 (9)	P22 (9)	–	P20(cont)
3	max(3,9)+3	0	0	P11 (6), P12 (9), P13(12)	P22 (9)	–	P11
4	max(4,12)+3	max(4,9)+6	0	P12 (9), P13(12) P14(15)	P22 (9), P24(15)	–	P12
5	max(5,15)+3	0	max(5,0)+3	P13(12), P14(15), P15(18)	P22 (9), P24(15)	P35 (8)	P35
6	0	max(6,15)+6	max(6,8)+3	P13(12), P14(15), P15(18)	P22 (9), P24(15)	P36 (11)	P22
7	0	0	max(7,11)+3	P13(12), P14(15), P15(18)	P24(15), P26(21)	P36(11), P37(14)	P22(cont)
8	0	max(8,21)+6	0	P13(12), P14(15), P15(18)	P24(15), P26(21)	P36(11), P37(14)	P36
9	0	0	0	P14(15), P15(18)	P24(15), P26(21)	P36(11), P37(14)	P13

Figure 9.17. *WFQ scheduling*

If a flow remains inactive while others are evolving, a new variable called virtual time is introduced. This value is incremented even in the absence of traffic.

When a packet arrives after a period of inactivity, the value of the timestamp corresponds to the sum of:

– the maximum value of the pair (virtual time incremented with each Ti, old timestamp);

– a value equal to the ratio of the length of the new packet to the weight.

The packet in queue QF3 is assigned the timestamp (8), corresponding to the sum of the virtual time (5) and the value (3) equal to the ratio of the length of the new packet to the weight.

PQ (Priority Queuing) scheduling defines a simple algorithm that consists of serving the higher priority queue as long as there are stored packets. When the highest priority queue becomes empty, the scheduler processes the packets in the following queue, and so on. When a packet arrives into a higher priority queue, it is served as soon as the output port finishes transmitting the current packet, regardless of its priority.

An overload in a higher priority class may starve lower priority classes. In order to avoid this drawback, the solution consists of limiting the bandwidth allocated to each priority queue.

10

Security Principles

10.1. Security services

Confidentiality consists of ensuring that data can only be read by authorized individuals. This service is obtained using a mechanism that encrypts the relevant data.

Integrity control consists of detecting modifications in transferred data. This service is obtained via a hash function or an encryption algorithm that generates a seal or a signature.

Authentication is to verify the origin of the data or the identity of the third party who wishes to access a network. This service is obtained from a seal or a signature.

Non-repudiation ensures that the data source is not able to contest operations it has performed.

This service is obtained from a signature. The seal may be calculated in two different ways:

– symmetric-encryption algorithm is applied to the data, the seal corresponding to the last block of cryptogram;

– seal is calculated from the encryption digest by a secret key.

Signature is calculated from the encryption digest by a private key and the decryption with a public key.

10.2. Encryption function

There are two major families of encryption algorithms: symmetric algorithms based on secret key and asymmetric algorithms based on public and private keys.

Symmetrical algorithms are grouped into two categories:

– stream cipher algorithms act on bits. A stream cipher generally consists of an exclusive OR or XOR (eXclusive OR) operation between data issued by a pseudo-random number generator and the data being transmitted;

– block cipher algorithms act on blocks of between 32 and 512 bits in size.

Symmetric algorithms are well suited to data encryption. Encryption is provided, for example, by the block cipher AES (Advanced Encryption Standard).

Asymmetrical algorithms are used mainly in the following two scenarios:

– Secret key transport. The secret key is encrypted by the public key and decrypted by the private key. Only the holder of the private key can recover the secret key;

– Signature. The data digest, calculated using a hash function, is encrypted by the private key and decrypted by the public key. The data source is the only holder of the private key, which guarantees the integrity control of the data received, the authentication of the data source and the non-repudiation by the client.

Encryption based on the secret key and the digital signature is provided, for example, by the RSA (named from the initials of the three developers Rivest, Shamir, Adleman) algorithm based on the modular exponentiation.

Asymmetrical algorithms are not adapted to data encryption because they have the disadvantage of being slow. This is due to the size of the keys used.

10.2.1. *AES algorithm*

The Advanced Encryption Standard (AES) is a symmetrical algorithm of block cipher. It operates on 128-bit data blocks and uses keys 128, 192 or 256 bits in length.

The 16-byte input data block is represented by a 4-line by 4-column table, with each cell containing a byte.

```
32 43 f6 a8 88 5a 30 8d 31 31 98 a2 e0 37 07 34
```

Input data

32	88	31	e0
43	5a	31	37
f6	30	98	07
a8	8d	a2	34

The 16-byte key is represented by a 4-line by 4-column the table, with each cell containing a byte.

```
2b 7e 15 16 28 ae d2 a6 ab f7 15 88 09 cf 4f 3c
```

Initial key

2b	28	ab	09
7e	ae	f7	cf
15	d2	15	4f
16	a6	88	3c

An exclusive OR is executed between the input data and the initial key. The output data makes up the entry data of the first iteration.

Input data					Initial key					Output data			
32	88	31	e0		2b	28	ab	09		19	a0	9a	e9
43	5a	31	37	$\oplus$	7e	ae	f7	cf	=	3d	f4	c6	f8
f6	30	98	07		15	d2	15	4f		e3	e2	8d	48
a8	8d	a2	34		16	a6	88	3c		be	2b	2a	08

The transformation `SubBytes` applies the S-Box substitution table to each byte of the input data table of the iteration.

For example, if the byte has a hexadecimal value of 19, the substitution value corresponds to the value of the first line and ninth column in the S-Box table.

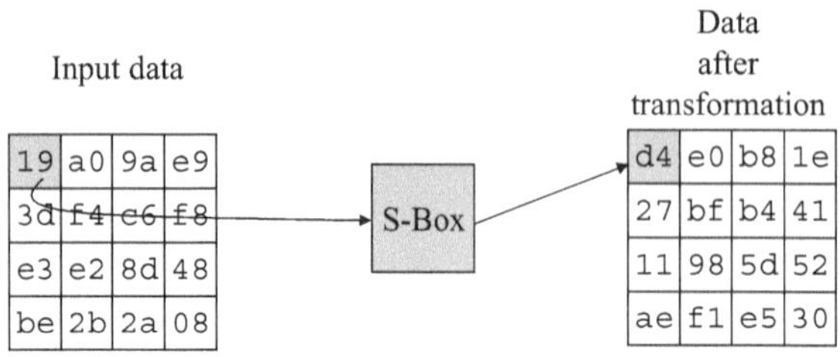

Iteration 1 : transformation `SubBytes`

The transformation `ShiftRows` consists of applying a shift to each line for the three last lines of the table resulting from the transformation `SubBytes`, with the first line being completely copied without modification. The second line undergoes a shift of 1 byte on the left, the third line a shift of 2 bytes and the fourth line a shift of 3 bytes.

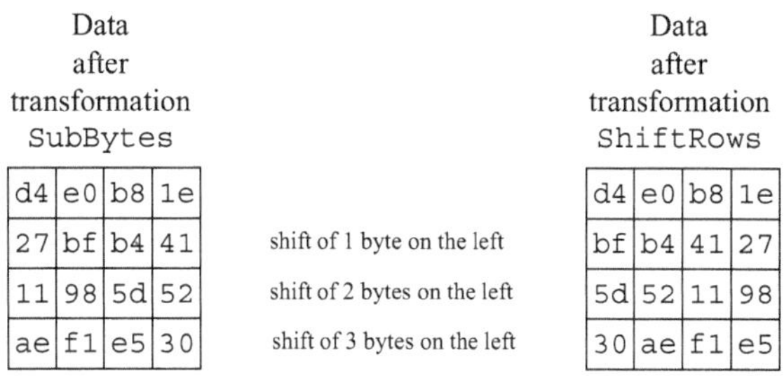

Iteration 1 : transformation `ShiftRows`

The transformation `MixColumns` operates on the columns of the table resulting from the transformation `ShiftRows`.

The transformation of the first column is executed as follows:

$$\{04\} = (\{02\} \bullet \{d4\}) \oplus (\{03\} \bullet \{bf\}) \oplus \{5d\} \oplus \{30\}$$

$$\{66\} = \{d4\} \oplus (\{02\} \bullet \{bf\}) \oplus (\{03\} \bullet \{5d\}) \oplus \{30\}$$

$$\{81\} = \{d4\} \oplus \{bf\} \oplus (\{02\} \bullet \{5d\}) \oplus (\{03\} \bullet \{30\})$$

$$\{e5\} = (\{03\} \bullet \{d4\}) \oplus \{bf\} \oplus \{5d\} \oplus (\{02\} \bullet \{30\})$$

The symbol • represents a multiplication of bytes in the form of polynomials, modulo the prime polynomial, $x^8 + x^4 + x^3 + x + 1$.

Data after transformation `ShiftRows`

d4	e0	b8	1e
bf	b4	41	27
5d	52	11	98
30	ae	f1	e5

Transformation matrix

02	03	01	01
01	02	03	01
01	01	02	03
03	01	01	02

Data after transformation `MixColumns`

04	e0	48	28
66	cb	f8	06
81	19	d3	26
e5	9a	7a	4c

Iteration 1 : transformation `MixColumns`

The first iteration is ended by an exclusive OR (`AddRoundKey` transformation) between the output data of the transformation `MixColumns` and the key of the first iteration. This key is derived from the initial key.

Data after transformation `MixColumns`

04	e0	48	28
66	cb	f8	06
81	19	d3	26
e5	9a	7a	4c

⊕

1st iteration key

a0	88	23	2a
fa	54	a3	6c
fe	2c	39	76
17	b1	39	05

=

Data after transformation `AddRoundKey`

a4	68	6b	02
9c	9f	5b	6a
7f	35	ea	50
f2	2b	43	49

Iteration 1 : transformation `AddRoundKey`

The output data of the first iteration constitutes the input data of the second iteration. The number of iterations is equal to 10 for a 128-bit key, 12 for a 192-bit key and 14 for a 256-bit key.

Key derivation starts with the cutting of the initial 128-bit key into four blocks, w_0, w_1, w_2 and w_3.

w_0 = 2b7e1516

w_1 = 28aed2a6

w_2 = abf71588

w3 = 09cf4f3c

Table 10.1 shows the derivation of the initial key for the first iteration.

i	4	5	6	7
w_{i-1}	09cf4f3c	a0fafe17	88542cb1	23a33939
After RotWord	cf4f3c09			
After SubWord	8a84eb01			
Rcon	01000000			
After exclusive OR	8b84eb01			
w_{i-4}	2b7e1516	28aed2a6	abf71588	09cf4f3c
w_i	a0fafe17	88542cb1	23a33939	2a6c7605

Table 10.1. *Derivation of the initial key for the first iteration*

The key of the first iteration is made up of four blocks w_4, w_5, w_6 and w_7. The transformation RotWord consists of the shift of a byte on the left. The transformation SubWord applies the substitution table S-Box to each byte. The block w_i is the result of an exclusive OR between blocks w_{i-1} (having possibly undergone transformations) and w_{i-4}.

10.2.2. *Secret key generation*

The Diffie–Hellman mechanism is used to generate a secret key based on modular exponentiation. The procedure occurs as follows:

– Alice and Bob choose an integer p such that $(p - 1)/2$ is prime and an integer g primitive with p;

– Alice chooses a secret number a such that $1 \leq a \leq p - 1$;

– Bob chooses a secret number b such that $1 \leq b \leq p - 1$;

– Alice calculates the value $A = g^a$ [mod p] and sends it to Bob;

– Bob calculates the value $B = g^b$ [mod p] and sends it to Alice;

– Alice generates the secret key equal to:

$$(B\ [\mathrm{mod}\ p])^a = B^a\ [\mathrm{mod}\ p] = (g^b)^a\ [\mathrm{mod}\ p] = g^{ba}\ [\mathrm{mod}\ p]$$

– Bob generates the same secret key equal to:

$$(A\ [\mathrm{mod}\ p])^b = A^b\ [\mathrm{mod}\ p] = (g^a)^b\ [\mathrm{mod}\ p] = g^{ab}\ [\mathrm{mod}\ p]$$

A third party can recover the values of p, g, $A = g^a$ [mod p] and $B = g^b$ [mod p], which circulate unencrypted in the network. This does not allow the third party to calculate g^{ab} [mod p] if the numbers (p, a, b) are large enough to avoid an attack through exhaustive search.

Table 10.2 supplies the formula used to calculate the value of the prime number p, with the value of g (generator) being fixed at 2, for MODP (MODular exponential modulus P) groups.

Group number	Designation	Value of p
1	768-bit MODP	$p = 2^{768} - 2^{704} - 1 + 2^{64} \times \{E[2^{638} \times pi] + 149{,}686\}$
2	1024-bit MODP	$p = 2^{1024} - 2^{960} - 1 + 2^{64} \times \{E[2^{894} \times pi] + 129{,}093\}$
5	1536-bit MODP	$p = 2^{1536} - 2^{1472} - 1 + 2^{64} \times \{E[2^{1406} \times pi] + 741{,}804\}$
14	2048-bit MODP	$p = 2^{2048} - 2^{1984} - 1 + 2^{64} \times \{E[2^{1918} \times pi] + 741{,}804\}$
15	3072-bit MODP	$p = 2^{3072} - 2^{3008} - 1 + 2^{64} \times \{E[2^{2942} \times pi] + 169{,}031\}$
16	4096-bit MODP	$p = 2^{4096} - 2^{4032} - 1 + 2^{64} \times \{E[2^{3966} \times pi] + 240{,}904\}$
17	6144-bit MODP	$p = 2^{6144} - 2^{6080} - 1 + 2^{64} \times \{E[2^{6014} \times pi] + 929{,}484\}$
18	8192-bit MODP	$p = 2^{8192} - 2^{8128} - 1 + 2^{64} \times \{E[2^{8062} \times pi] + 4{,}743{,}158\}$

Table 10.2. *Values of p for MODP groups*

10.2.3. *RSA algorithm*

Alice wants to receive a block of confidential data M from Bob. Alice generates a public key that she transmits to Bob and a private key that she keeps. Bob encrypts the data block M with the public key. Only Alice can recover the data block M using her private key.

RSA algorithm is implemented using the following procedure:

– Alice generates two prime numbers, p and q: $p = 59$ and $q = 71$;

– Alice calculates the product $w = (p - 1)(q-1)$: $w = 58 \times 70 = 4{,}060$;

– Alice chooses a prime number e with w: $e = 671$. It is relatively easy to determine the value e using Euclid's algorithm to verify whether it is prime or not with w;

– Alice calculates the product $n = p \times q$: $n = 59 \times 71 = 4{,}189$;

– Alice calculates the number d such that $d \times e \equiv 1$ [mod w]: $d = 1{,}791$. The calculation of the value d uses an extended Euclidean algorithm or Bézout algorithm;

– Alice transmits to Bob the values n and e that constitute the public key, and keeps the values n and d that represent the private key;

– Bob encrypts his message M with the public key: $M \mapsto M^{671}$ [mod 4189];

$M = 01010010\ 01010011\ 01000001$

M is cut into blocks of length k such that $2^k < n < 2^{k+1}$

M is therefore cut into blocks of 12 bits: M = 010100100101 001101000001

The twelve bits 010100100101 converted into decimals give 1317

The twelve bits 001101000001 converted into decimals give 833

The encrypted message C transmitted by Bob is as follows:

- Bob calculates 1317^{671} [mod 4189], or 3530
- Bob calculates 833^{671} [mod 4189], or 3050

– Alice decrypts the received message C with her private key:

$C \mapsto C^{1791}$ [mod 4189]

- Alice calculates 3530^{1791} [mod 4189] and recovers the value 1317
- Alice calculates 3050^{1791} [mod 4189] and recovers the value 833

The difficulty of the RSA algorithm lies in the determination by Alice of two prime numbers, p and q, and in the calculation of M^e [mod n] and C^d [mod n]. The generation of large prime numbers constitutes a relatively complex problem.

10.2.4. *Public key distribution*

Asymmetrical encryption algorithms are based on the diffusion of the public key by the holder of the private key. Conversely, nothing guarantees the identity of the holder of the private key.

The certificate, signed by a trusted authority, enables a public key to be associated with its holder. All the recipients of the public key possess the public key of the trusted authority, which enables them to ensure the authenticity of the certificate.

Certificates are managed by a Public Key Infrastructure (PKI), which ensures the following functions:

- user registration;
- generation of certificates;
- renewal of certificates;
- revocation of certificates;
- publication of certificates;
- publication of revocation lists.

The public key infrastructure is divided into several entities:

– the Certification Authority (CA), whose mission is to sign Certificate Signing Requests (CSR) and Certificate Revocation Lists (CRL);

– the Registration Authority (RA), whose mission is to generate certificates and to conduct usage verifications of the identity of the final user;

– the repository authority, whose mission is to store digital certificates and revocation lists;

– the key escrow authority, which provides authorities with the means of decrypting data for a user.

A certificate is composed of two parts:

– the data being certified, which contains the base fields and extensions;

– the certificate signature.

The base fields of a certificate provide the following information:

– the structure version of the certificate (value = 3);

– the serial number;

– the information about the certificate signature (algorithms and parameters);

– the name of the certificate issuer;

– the validity period of the certificate;

– the name of the certificate holder;

– the public key (value of the public key, algorithm and parameters).

Certificate extensions enable a more precise specification of the following characteristics:

– information about keys;

– information on certification policies;

– additional information on the issuer and holder of the certificate;

– limitations on the certification process.

10.3. Hash function

The hash function is the mechanism that converts a string of bytes (the data being transmitted) into a smaller string called a digest. When the data to be transmitted is associated with a secret key, the hash function enables the sealing of these data to provide authentication of the data source and integrity control of the data. The non-repudiation service is not provided.

10.3.1. *MD5 algorithm*

The Message Digest (MD5) algorithm calculates a 128-bit digest from a block of data cut into 512-bit data blocks.

The first stage consists of filling the data block. The data block is completed so as to be congruent to 448 bits modulo 512. Filling is always executed, even if the message is already congruent to 448 bits modulo 512. The filling value therefore varies from 1 to 512 bits. Filling is executed as follows:

– the final filling bit is set at ONE;

– the other filling bits are set at ZERO.

The length of the data block is coded on 64 bits. This field is added to the structure previously defined so as to obtain 512-bit data blocks.

Calculation of the digest is based on registers A, B, C, and D, which are 32 bits in length. For the first 512-bit block, these registers are initialized with the following values (initialization vector) in hexadecimals (Figure 10.1):

A = 67452301

B = EFCDAB89

C = 98BADCFE

D = 10325476

For the following 512-bit blocks, these registers are initialized from the values obtained for the previous 512-bit block. When all of the 512-bit blocks have been processed, the final digest corresponds to the concatenation of the values of registers A, B, C, and D (Figure 10.1).

Operations on a 512-bit block are divided into four stages, which are themselves subdivided into 16 elementary iterations based on a function that varies according to stage, on an addition and a shift to the left (Figure 10.2).

When the four stages are finished, the values of registers A, B, C, and D are added to those of the initialization vector ((Figure 10.1).

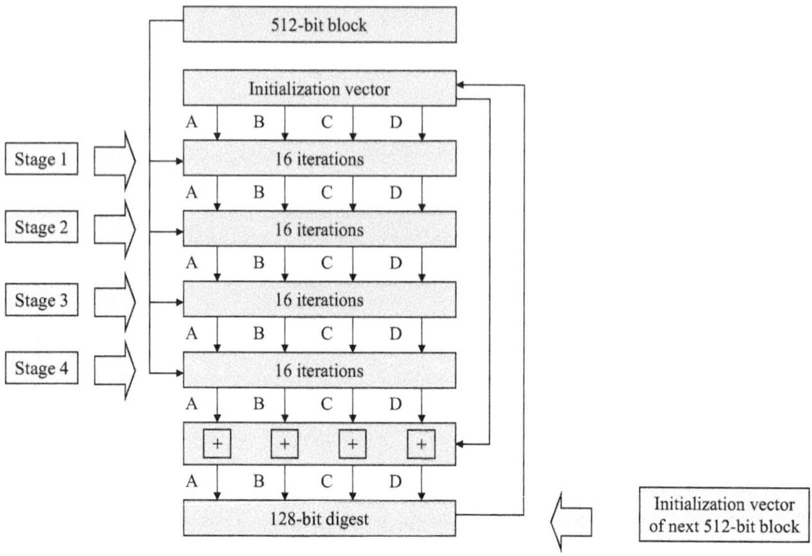

Note: The + sign symbolizes an addition of modulo 2^{32}.

Figure 10.1. *MD5 algorithm*

The four functions defined are:

Stage one: F(X, Y, Z) = (X AND Y) OR ((NO X) AND Z)

Stage two: F(X, Y, Z) = (X AND Y) OR (Y AND (NO Z))

Stage three: F(X, Y, Z) = X XOR Y XOR Z

Stage four: F(X, Y, Z) = Y XOR (X OR (NO Z))

The elementary iteration is shown in Figure 10.2. The new values of registers A, B, C, and D are supplied, for stage one, by the following equations:

new A = previous D

new B = ((previous A + F (previous B, previous C, previous D) + k[i] + w[g]) <<< r[i]) + previous B.

new C = previous B

new D = previous C

Each 512-bit data block is subdivided into 16 words w[i] of 32 bits with $0 \leq i \leq 15$. The rank of the stage determines the index value [g]:

g = i, with $0 \leq i \leq 15$ for stage one

g = (5×i + 1) [mod 16], with $16 \leq i \leq 31$ for stage two

g = (3×i + 5) [mod 16], with $32 \leq i \leq 47$ for stage three

g = (7×i) [mod 16], with $48 \leq i \leq 63$ for stage four

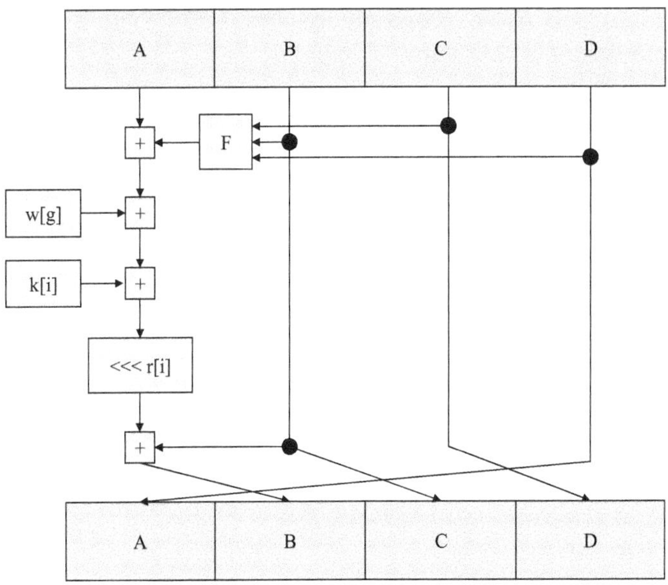

Figure 10.2. *Elementary iteration of MD5 algorithm*

The symbol <<< indicates that a rotation to the left of r[i] bits must be executed. The value of r[i] is supplied by the following table for each stage:

r[0..15] = {7, 12, 17, 22, 7, 12, 17, 22, 7, 12, 17, 22, 7, 12, 17, 22}

r[16..31] = {5, 9, 14, 20, 5, 9, 14, 20, 5, 9, 14, 20, 5, 9, 14, 20} r[32..47] = {4, 11, 16, 23, 4, 11, 16, 23, 4, 11, 16, 23, 4, 11, 16, 23}

r[48..63] = {6, 10, 15, 21, 6, 10, 15, 21, 6, 10, 15, 21, 6, 10, 15, 21}

The value of k[i] is determined by the following formula:

k[i] = E(abs(sin(i + 1)) × 2^{32}), where E designates the integer part and (i+1) is expressed in radians.

10.3.2. *SHA algorithm*

The acronym SHA (Secure Hash Algorithm) is associated with multiple hash algorithms: SHA-0, SHA-1, and SHA-2.

Algorithm SHA-0 is inspired by the MD5 algorithm. It is not recommended for use for security questions.

Algorithm SHA-1 is a modified version of algorithm SHA-0.

Algorithm SHA-2 is composed of two algorithms:

– algorithm SHA-256, including versions SHA-224 and SHA-256;

– algorithm SHA-512, including versions SHA-384, SHA-512, SHA-512/256 and SHA-512/224.

10.3.2.1. *SHA-1 algorithm*

The operation of the SHA-1 algorithm is similar to that of the MD5 algorithm, for which the principal differences introduced are:

– there are five registers (A, B, C, D and E) used to produce the 160-bit digest;

– the number of elementary iterations increases to 80.

Figure 10.3 shows the elementary iteration of the SHA-1 algorithm.

The SHA-1 algorithm uses a Boolean function F among four and changes function every 20 iterations. Likewise, the value of k[i] remains constant for 20 iterations.

The number of words w[g] is equal to 80, one word per iteration. The first 16 words belong to the 512-bit block. The next 64 words are obtained using a rotation applied to the result of an exclusive OR of four words calculated during previous iterations.

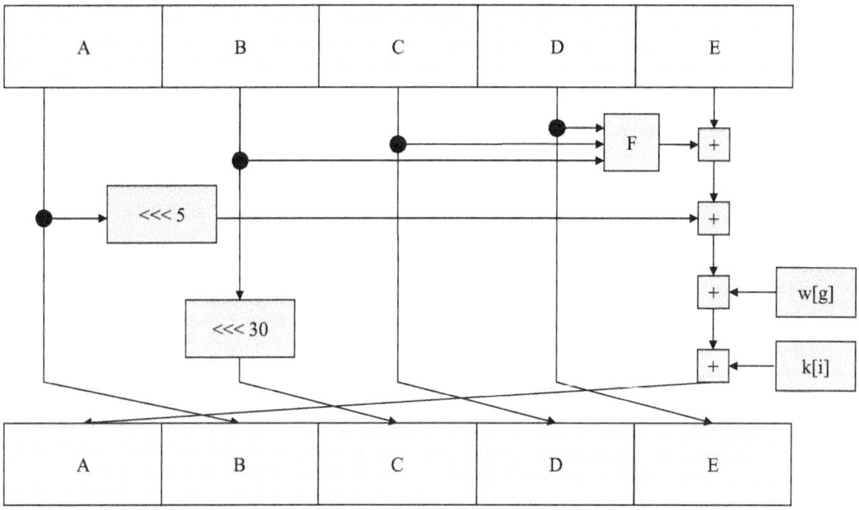

Figure 10.3. *Elementary iteration of SHA-1 algorithm*

10.3.2.2. *SHA-2 algorithm*

Algorithm SHA-226 and its truncated version, SHA-224, are structured on the basis of 32-bit words and a cutting into 512-bit blocks.

Algorithm SHA-512 and its truncated versions, SHA-384, SHA-512/256, and SHA-512/224, use 64-bit words and a cutting into 1,024-bit blocks.

The SHA-2 algorithm has the following specific characteristics:

– there are 8 registers (A, B, C, D, E, F, G and H). The capacity of each register is 32 bits (as for SHA-1 or MD5) in the case of the SHA-256 algorithm, or 64 bits in the case of the SHA-512 algorithm.

– there are 64 elementary iterations in the case of SHA-256 (as for MD5) and 80 in the case of SHA-512 (as for SHA-1);

– the length of the data block is coded at the end of filling on 64 bits in the case of SHA-256 (as for SHA-1 or MD5) and on 128 bits in the case of SHA-512.

The digest of each version is indicated by the last suffix of the algorithm concerned:

– the SHA-256 algorithm produces a 256-bit digest generated by the content of eight registers (A, B, C, D, E, F, G and H);

– the SHA-224 algorithm uses the SHA-256 algorithm and produces a 224-bit digest generated by the content of seven registers (A, B, C, D, E, F and G);

– the SHA-512 algorithm produces a 512-bit digest generated by the content of eight registers (A, B, C, D, E, F, G and H);

– the SHA-384 algorithm uses the SHA-512 algorithm and produces a 384-bit digest generated by the content of six registers (A, B, C, D, E and F);

– the SHA-512/256 algorithm uses the SHA-512 algorithm and produces a 256-bit digest generated by the content of four registers (A, B, C and D);

– the SHA-512/224 algorithm uses the SHA-512 algorithm and produces a 224-bit digest generated by the content of three registers (A, B, C and C) and the first 32 bits of register D.

The operating principle relative to each iteration is identical for all of the algorithms in the SHA-2 family. However, these algorithms differ in the following ways:

– SHA-256 and SHA-224 algorithms use a specific initialization vector;

– SHA-512, SHA-384 and SHA-512/256 algorithms use a specific initialization vector;

– SHA-256 and SHA-224 algorithms, on the one hand, and the SHA-512, SHA-384 and SHA-512/256 algorithms, on the other hand, use Boolean functions (Ch, Ma), rotation operations ($\Sigma 0$, $\Sigma 1$) and specific k[i] constants. Likewise, the calculation of the word w[g] is specific for $g \geq 16$.

Figure 10.4 shows the elementary iteration of the SHA-2 algorithm.

10.3.3. *HMAC function*

The HMAC (Keyed-Hash Message Authentication Code) mechanism defines the association of secret key K and a data block D used for the calculation of the MAC (Message Authentication Code) using a hash function. The digest obtained, which acts as a seal, enables the checking of the authenticity (origin and integrity) of data block D.

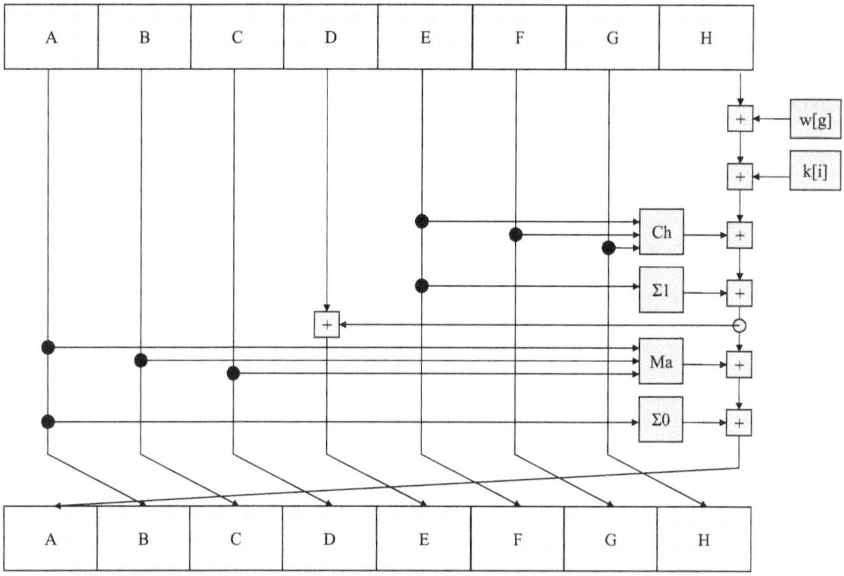

Figure 10.4. *Elementary iteration of SHA21 algorithm*

The HMAC mechanism is also used to calculate derived keys.

The seal is calculated using the formula below:

$$MAC(D) = HMAC(K, D) = H((K0 \oplus opad) || H((K0 \oplus ipad) || D))$$

where:

– H is the hash function (MD5 or SHA);

– K0 is the key calculated from secret key K;

– opad (outer padding) is the filling composed of byte 5C (in hexadecimals) repeated B times;

– ipad (inner padding) is the filling composed of bite 36 (in hexadecimals) repeated B times;

– B is the size of the data block to which the hash function is applied. If the block size of the hash function is 512 bits, ipad and opad are 64 repetitions of bytes 5C and 36 (in hexadecimals).

The sign || designates the concatenation of bytes.

The calculation of the MAC seal takes place in the following stages:

– stage 1: the length of key K is equal to B

K0 = K

go to stage 4;

– stage 2: the length of key K is greater than B

apply the hash function H to key K to obtain a digest of length L

complete the digest calculated with ZEROS to obtain size B

K0 = H(K) || 00....00

go to stage 4;

– stage 3: the length of key K is smaller than B

complete key K with ZEROS to obtain size B;

– stage 4: execute an exclusive OR between key K0 and ipad to obtain the structure K0 $\oplus$ ipad;

– stage 5: concatenate data D to authenticate with the structure generated in stage 4 to obtain the structure K0 $\oplus$ ipad) || D;

– stage 6: apply hash function H to the structure generated in stage 5 to obtain the structure H((K0 $\oplus$ ipad) || D);

– stage 7: execute an exclusive OR between key K0 and opad to obtain the structure K0 $\oplus$ opad;

– stage 8: concatenate the structures generated in stages 6 and 7 to obtain the structure (K0 $\oplus$ opad) || H((K0 $\oplus$ ipad) || D);

– stage 9: apply hash function H to the structure generated in stage 8 to obtain the seal enabling authentication and integrity control of data D.

It is possible to truncate the structure generated in stage 9 to retain a reduced number of left t bits as a seal. For security questions, the number *t* must be greater than or equal to max(L/2, 80), with L being the length of the digest. The seal is designated in the form of HMAC-H-t. For example, notation HMAC-SHA1-80 designates the calculation of a seal using hash function SHA-1, the digest of which is truncated to 80 bits.

11

Security Mobile Network

11.1. Security architecture

The security architecture implemented in the 4G mobile network is dependent on the execution of the following:

– mutual authentication between User Equipment (UE) and Evolved Packet System (EPS);

– the security of the signaling messages NAS (Non-Access Stratum) exchanged between the mobile and the Mobility Management Entity (MME). The security services concerns the integrity control and encryption of the NAS messages;

– the security of the radio interface LTE-Uu between the mobile and the evolved Node Base station (eNB). The security services regards, firstly, the integrity control and the encryption of the messages RRC (Radio Resource Control), and, secondly, the encryption of the IP (Internet Protocol) packets belonging to the user plane.

The authentication is based on AUTN and RES seals generated by the HSS entity and the mobile from a random sequence RAND and the secret key Ki. The RAND sequence is generated by the Home Subscriber Server (HSS), and then transmitted to the mobile. The secret key Ki is generated when creating the subscription (Figure 11.1).

The security of the NAS messages comes from deriving the K_{ASME} key to generate the integrity control key K_{NASint} and the encryption key K_{NASenc}. The

K_{ASME} key is generated by the HSS entity and the mobile from the RAND sequence and the secret key Ki. The derivation of the K_{ASME} key is achieved, firstly by the MME and, secondly, by the mobile (Figure 11.1).

The security of the radio interface LTE-Uu come from deriving the K_{eNB} key to generate, on the one hand, the integrity control key K_{RRCint} and the encryption key K_{RRCenc} for RRC messages, and, on the other hand, the encryption key K_{UPenc} for the IP packets. Deriving thc K_{eNB} key is achieved, firstly, by the eNB entity, and, secondly, by the mobile (Figure 11.1).

The K_{eNB} key is derived from the K_{ASME} key. The derivation of the K_{ASME} key is achieved, firstly, by the MME and, secondly, by the mobile (Figure 11.1)

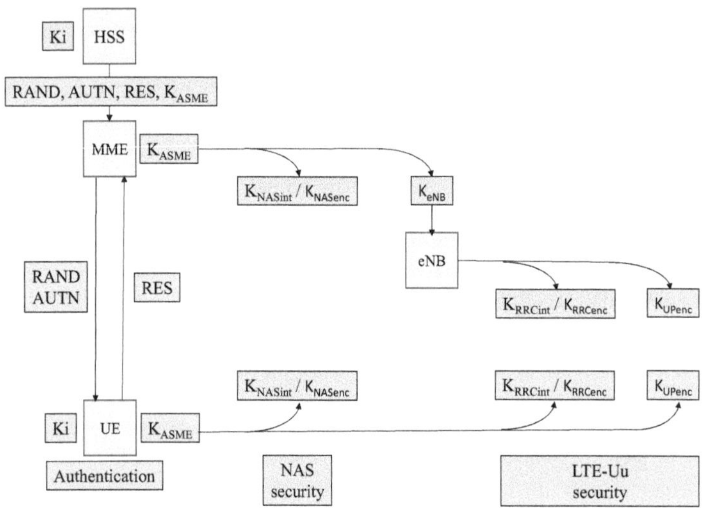

Figure 11.1. *Security architecture*

11.2. Mutual authentication

The mutual authentication procedure is described in Figure 11.2.

1) The attachment procedure is initiated by the mobile when it sends to the MME entity the message EMM ATTACH REQUEST containing the following information:

– International Mobile Subscriber Identity (IMSI);

– EPS Encryption Algorithm (EEA) and EPS Integrity Algorithm (EIA) for the implementation of the NAS message security and radio interface LTE-Uu available at the mobile (Table 11.1);

– the KSI_{ASME} parameter whose value is equal to 7 which indicates that the mobile does not have any K_{ASME} key.

Encryption algorithm EEA		Integrity control algorithm EIA	
EEA0	Null encryption algorithm	EIA0 (note)	Null integrity control algorithm
128-EEA1	SNOW 3G	128-EIA1	SNOW 3G
128-EEA2	AES	128-EIA2	AES
128-EEA3	ZUC	128-EIA3	ZUC

Note: EIA0 is allowed in an unauthorized emergency call only.

Table 11.1. *Algorithm for encryption and integrity control*

2) Upon receipt of the message EMM ATTACH REQUEST, the MME entity requests the HSS entity for the cryptographic data concerning the mobile in the message DIAMETER AIR (Authentication-Information-Request) containing the following information:

– the private identity IMSI of the mobile;

– the SN ID (Serving Network Identifier) parameter containing the Mobile Country Code (MCC) and the Mobile Network Code (MNC);

– the number of authentication vectors (RAND, AUTN, RES, K_{ASME});

– the type of radio access network, e.g. Evolved Universal Terrestrial Radio (E-UTRAN).

The HSS entity generates the RAND sequence, integrating SQN sequence number, and creates the parameters RES, AUTN, CK and IK using the AKA (Authentication and Key Agreement) algorithm from the key Ki and RAND sequence.

The HSS entity uses the Key Derivation Function (KDF) for generating the K_{ASME} key from the CK, IK, SQN and SN-ID parameters.

3) The HSS entity transmits the authentication vectors to the MME entity in the message DIAMETER AIA (Authentication-Information-Answer).

4) The MME entity transmits to the mobile the message EMM AUTHENTICATION REQUEST containing the random number RAND, the AUTN seal of the network and the KSI_{ASME} parameter that represents the index of the K_{ASME} key.

The mobile calculates locally, from its key Ki stored in the Universal Services Identity Module (USIM) of its Universal Integrated Circuit Card (UICC) and the RAND number received, the K_{ASME} key, its RES seal and that of the network AUTN. The mobile compares the received value AUTN with that calculated locally. If both values are identical, the network is authenticated.

5) The mobile responds to the MME entity with the message EMM AUTHENTICATION RESPONSE containing the seal RES.

The MME entity compares the received values RES from the mobile and the HSS entity. If the values are identical, the mobile is authenticated.

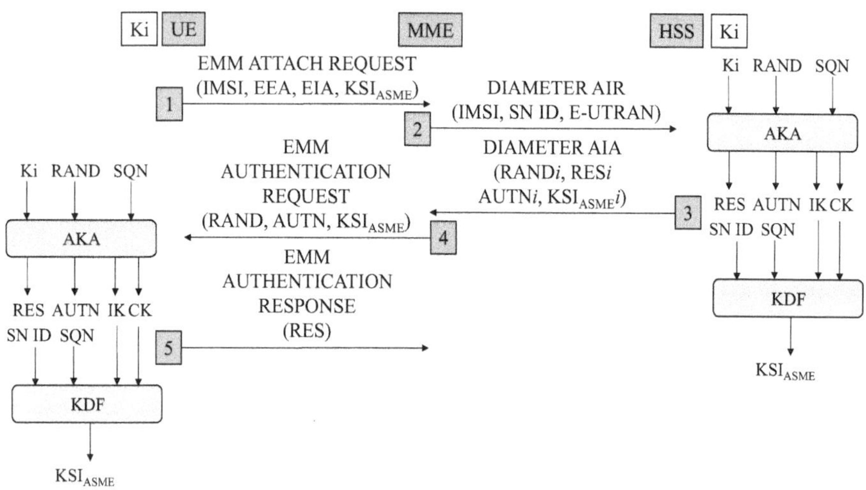

Figure 11.2. *Mutual authentication procedure*

11.3. NAS security

The MME entity derives the K_{ASME} key to generate the integrity control key K_{NASint} and the encryption key K_{NASenc} using the algorithm identifier (Alg-ID) and the algorithm distinguisher (NAS-int-alg, NAS-enc-alg) (Table 11.2).

$$K_{NASint} = KDF(K_{ASME}, \text{NAS-int-alg}, \text{Alg-ID})$$

$$K_{NASenc} = KDF(K_{ASME}, \text{NAS-enc-alg}, \text{Alg-ID})$$

Identifier	Description	Value
EEA0	Null algorithm	0000
128-EEA1	SNOW 3G	0001
128-EEA2	AES	0010
128-EEA3	ZUC	0011
128-EIA1	SNOW 3G	0001
128-EIA2	AES	0010
128-EIA3	ZUC	0011

Distinguisher	Value
NAS-int-alg	(0x)01
NAS-enc-alg	(0x)02
RRC-int-alg	(0x)03
RRC-enc-alg	(0x)04
UP-enc-alg	(0x)05

Table 11.2. *Algorithm identifiers and distinguishers*

The security set-up procedure of NAS messages is described in Figure 11.3.

1) The MME entity transmits to the mobile the message EMM SECURITY MODE COMMAND, controlled in integrity with the K_{NASint} key. The message contains the algorithm for integrity control and encryption selected by the MME entity.

The MAC (Message Authentication Code) seal is calculated with the integrity check algorithm EIA selected by the MME using the following parameters:

– a 32-bit counter (count) of the NAS messages;

– the direction of the message, encoded on a bit, ZERO for the uplink, ONE for the downlink;

– the identifier of the bearer, encoded on 5 bits, whose value is fixed (all bits to ZERO).

The mobile derived the K_{ASME} key to generate the integrity control key K_{NASint} and the encryption key K_{NASenc} using the algorithm (Alg-ID) and the algorithm distinguisher (NAS-int-alg, NAS -enc-alg).

The mobile calculates locally the MAC seal of the message EMM SECURITY MODE COMMAND and compares it with the MAC seal received from the MME entity. If the two values are equal, the message EMM SECURITY MODE COMMAND is accepted.

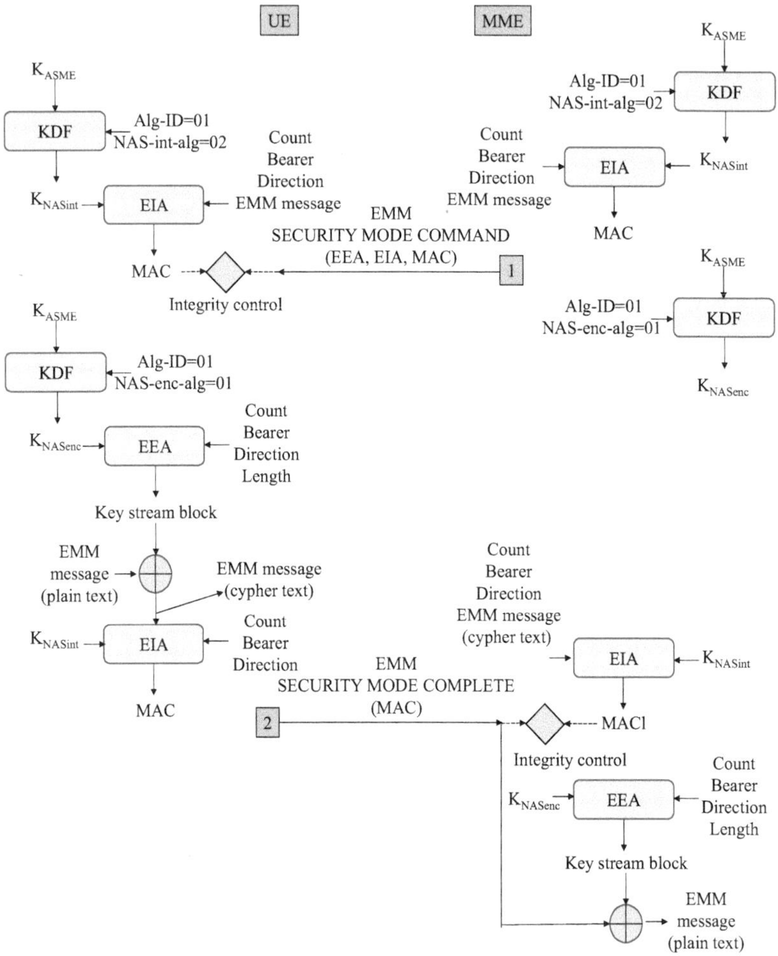

Figure 11.3. *NAS security*

2) The mobile responds with the message EMM SECURITY MODE COMPLETE, which is encrypted with the K_{NASenc} key and controlled in integrity with the K_{NASint} key.

The MME entity calculates locally the MAC seal of the message EMM SECURITY MODE COMPLETE and compares it with the MAC seal received from the mobile. If the two values are equal, the message EMM SECURITY MODE COMPLETE is accepted and the MME entity decrypts the message EMM COMPLETE SECURITY MODE.

11.4. LTE-Uu security

The radio interface LTE-Uu security set-up procedure is described in Figure 11.4.

The MME derives the K_{ASME} key to generate the K_{eNB} key using the counter UL-NAS COUNT of the NAS messages for the uplink direction.

1) The MME entity responds to the mobile with the message EMM ATTACH ACCEPT carried by the message S1-AP INITIAL CONTEXT SETUP REQUEST on the S1-MME interface.

The message S1-AP INITIAL CONTEXT SETUP REQUEST contains the K_{eNB} key derived from the K_{ASME} key and the algorithms for integrity control and encryption supported by the mobile.

The eNB entity derives the K_{eNB} key to generate the integrity control key K_{RRCint}, the encryption key K_{RRCenc} and the K_{UPenc} key for traffic encryption using the algorithm identifier (Alg-ID) and the algorithm distinguisher (RRC-int-alg, RRC-enc-alg, UP-enc-alg) (Table 11.2).

$$K_{RRCint} = KDF(K_{eNB}, \text{RRC-int-alg}, \text{Alg-ID})$$

$$K_{RRCenc} = KDF(K_{eNB}, \text{RRC-enc-alg}, \text{Alg-ID})$$

$$K_{UPenc} = KDF(K_{eNB}, \text{UP-enc-alg}, \text{Alg-ID})$$

2) The eNB entity requests the mobile making the security of the radio interface LTE-Uu with the message RRC *SecurityModeCommand* that is controlled in integrity with the K_{RRCint} key and contains algorithms that allow the mobile to derive the K_{eNB} key.

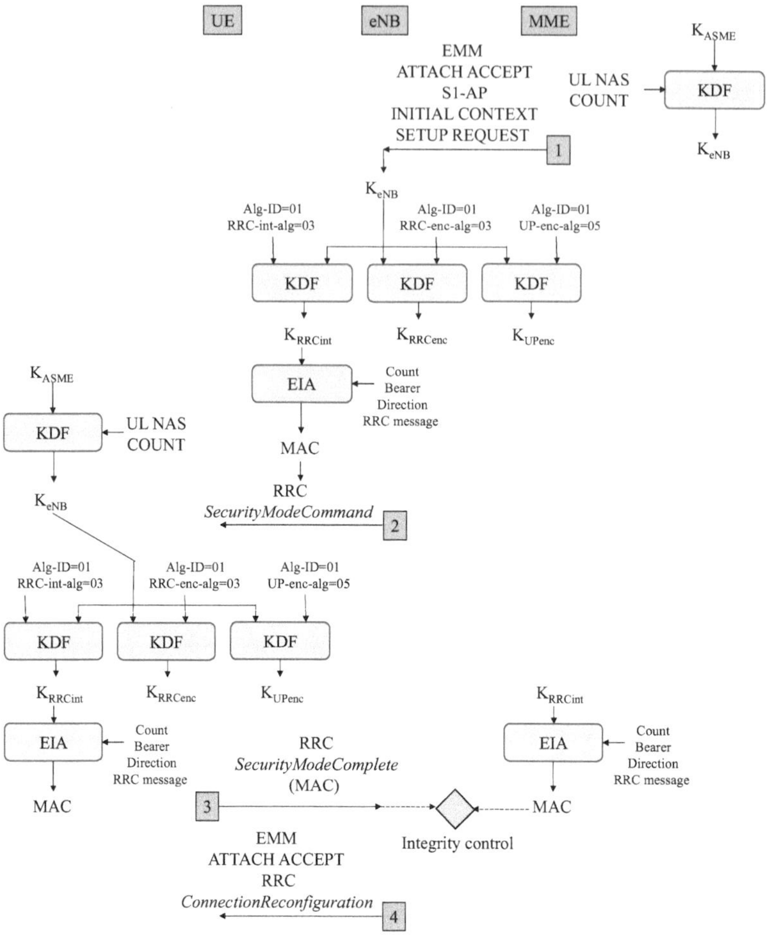

Figure 11.4. *LTE-Uu security*

The MAC seal is calculated with the integrity control algorithm EIA used by the eNB entity using the following parameters:

– a 32-bit counter (count) of the PDCP (Packet Data Convergence Protocol) frame;

– the direction of the message, encoded on a bit, ZERO for the uplink, ONE for the downlink;

– the identifier of the bearer, encoded on 5 bits, whose value is fixed (all bits to ZERO).

The mobile derived the K_{ASME} key to generate the K_{eNB} key, and then derives the K_{eNB} key to generate the integrity control key K_{RRCint}, the encryption key K_{RRCenc} and the K_{UPenc} key for the traffic encryption.

The mobile calculates locally the MAC seal of the message RRC *SecurityModeCommand* and compares it with the MAC seal received from the eNB cntity. If the two values are equal, the message RRC *SecurityModeCommand* is accepted.

3) The mobile confirms to the eNB entity the establishment of the keys with the message RRC *SecurityModeComplete*, which is controlled in integrity with the K_{RRCint} key.

The eNB entity calculates locally the MAC seal of the message RRC *SecurityModeComplete* and compares it with the MAC seal received from the mobile. If the two values are equal, the message RRC *SecurityModeComplete* is accepted.

4) The eNB entity transfers to the mobile the message EMM ATTACH ACCEPT message carried by RRC *ConnectionReconfiguration* on the radio interface LTE-Uu when the security of the radio interface LTE-Uu is effective.

11.5. Security context

Tables 11.3–11.6 summarize the parameters for the security context stored in the different entities.

Parameters	Source
IMSI	Configuration
SN ID	DIAMETER AIR (from MME)
Ki	Configuration
RAND, SQN	Calculation
AUTN	Calculation
RES	Calculation
K_{ASME}	Calculation

Table 11.3. *Security context at HSS entity*

Parameters	Source
IMSI	EMM ATTACH REQUEST (from UE)
SN ID	Configuration
Mobile capacities	EMM ATTACH REQUEST (from UE)
RAND	HSS DIAMETER AIA
AUTN	HSS DIAMETER AIA
RES	DIAMETER AIA (from HSS) EMM AUTHENTICATION RESPONSE (from UE)
K_{ASME}	DIAMETER AIA (from HSS)
KSI_{ASME}	MME
NAS algorithms	MME
K_{NASint}	Calculation
K_{NASenc}	Calculation
K_{eNB}	Calculation

Table 11.4. *Context security at MME entity*

Parameters	Source
IMSI	Configuration
Ki	Configuration
Mobile capacities	Configuration
RAND, SQN	EMM AUTHENTICATION REQUEST (from MME)
AUTN	EMM AUTHENTICATION REQUEST (from MME) and calculation
RES	Calculation
K_{ASME}	Calculation
KSI_{ASME}	EMM AUTHENTICATION REQUEST (from MME)
NAS algorithms	EMM SECURITY MODE COMMAND (from MME)
K_{NASint}	Calculation
K_{NASenc}	Calculation
K_{eNB}	Calculation
LTE-Uu algorithms	(from eNB) RRC *SecurityModeCommand*
K_{RRCint}	Calculation
K_{RRCenc}	Calculation
K_{UPenc}	Calculation

Table 11.5. *Mobile context security*

Parameters	Source
Mobile capacities	S1-AP INITIAL CONTEXT SETUP REQUEST (from MME)
K_{eNB}	S1-AP INITIAL CONTEXT SETUP REQUEST (from MME)
LTE-Uu algorithms	eNB
K_{RRCint}	Calculation
K_{RRCenc}	Calculation
K_{UPenc}	Calculation

Table 11.6. *Context security at eNB entity*

12

Security Transport Network

12.1. IPSec architecture

Security services (authentication, integrity and confidentiality) are offered in an identical way in IPv4 and IPv6. Their implementation is optional in IPv4 and mandatory in IPv6. Their use is optional.

The Internet Protocol Security (IPSec) mechanism can be used with Local Area Network (LAN) hosts or Security Gateways (SEGs) located at the interface between the LAN and Wide Area Network (WAN) networks.

The first application consists of linking remote LAN networks via an untrusted WAN network. The IPSec mechanism is deployed in the SEGs of each LAN network. In this case, the two LAN networks are considered as protected networks.

This configuration is deployed for connecting the entities of the 4G core network. The IPsec mechanism allows the protection of IP flows (S5 bearer) and GPRS Tunnel Protocol version 2 Control (GTPv2-C) messages (Figure 12.1).

The second application consists of linking a host to a remote LAN network via an untrusted WAN network. The IPSec mechanism is deployed in the host and a SEG located in the remote LAN network considered as a protected network.

This configuration is deployed to connect the evolved Node Base (eNB) station to the Mobility Management Entity and Serving Gateway of the 4G core network. The IPsec mechanism allows the protection of IP flows (S1 bearer) and S1-AP messages (Figure 12.2).

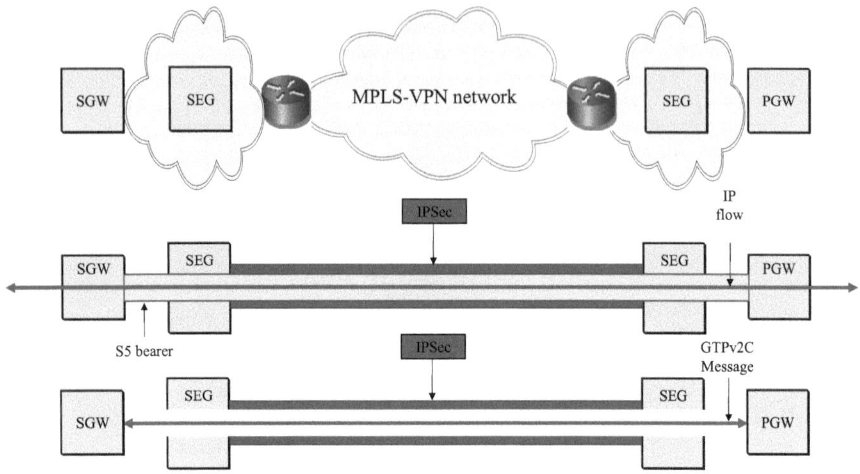

Figure 12.1. *IPSec Architecture*
Connecting entities of the 4G core network

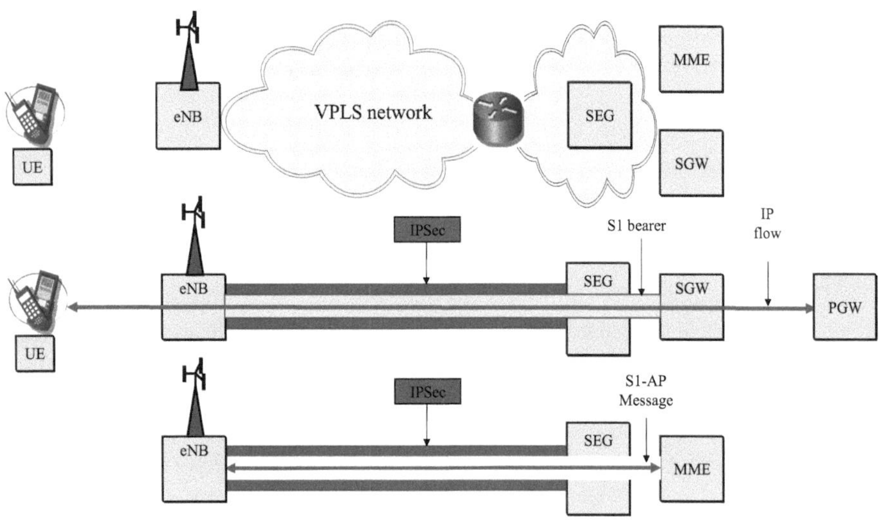

Figure 12.2. *IPSec Architecture*
Connecting eNB entity to MME and SGW entities

The third application consists of linking two hosts via two LAN networks and an untrusted WAN network. The IPSec mechanism is deployed from end to end in each host.

This configuration is deployed for the connection between eNB entities. The IPsec mechanism provides protection for incoming IP traffic (X2 bearer) and X2-AP messages (Figure 12.3).

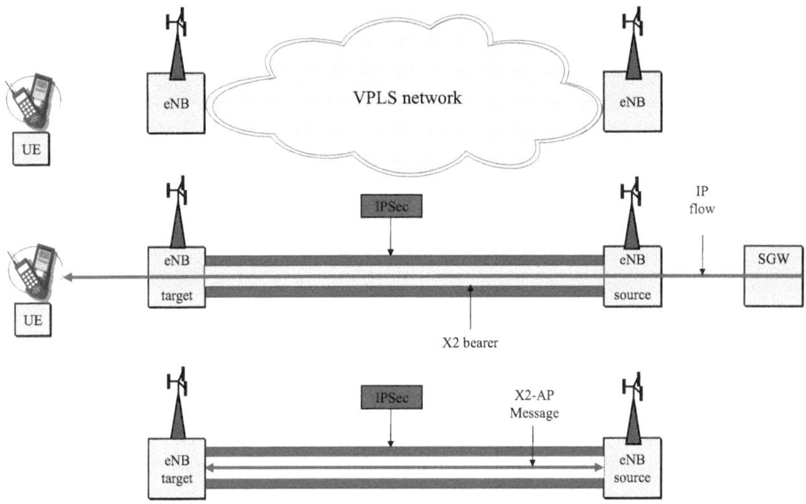

Figure 12.3. *IPSec architecture Connecting eNB entities*

12.1.1. *Security headers*

IPSec mechanism introduces two IPv4 or IPv6 header extensions:

– Authentication Header (AH) is designed to ensure the integrity and authentication of IP packets without data encryption (no confidentiality);

– Encapsulating Security Payload (ESP) ensures the confidentiality of IP packets, and optionally integrity and authentication.

12.1.1.1. *AH extension*

The AH extension offers authentication and data integrity services and enables protection against IP packet replay. The same AH extension is used in association with an IPv4 or IPv6 header.

The presence of the extension is indicated by the Next Header (in IPv6) or Protocol (in IPv4) field of the previous header with a value of 51.

In addition to the Next Header field, the AH extension contains the following fields (Figure 12.4):

```
0                   1                   2                   3
0 1 2 3 4 5 6 7 8 9 0 1 2 3 4 5 6 7 8 9 0 1 2 3 4 5 6 7 8 9 0 1
+-+-+-+-+-+-+-+-+-+-+-+-+-+-+-+-+-+-+-+-+-+-+-+-+-+-+-+-+-+-+-+-+
| Next Header |  Payload Len  |              RESERVED             |
+-+-+-+-+-+-+-+-+-+-+-+-+-+-+-+-+-+-+-+-+-+-+-+-+-+-+-+-+-+-+-+-+
|                 Security Parameters Index (SPI)                 |
+-+-+-+-+-+-+-+-+-+-+-+-+-+-+-+-+-+-+-+-+-+-+-+-+-+-+-+-+-+-+-+-+
|                     Sequence Number Field                       |
+-+-+-+-+-+-+-+-+-+-+-+-+-+-+-+-+-+-+-+-+-+-+-+-+-+-+-+-+-+-+-+-+
|                                                                 |
+                Integrity Check Value-ICV (variable)             |
|                                                                 |
+-+-+-+-+-+-+-+-+-+-+-+-+-+-+-+-+-+-+-+-+-+-+-+-+-+-+-+-+-+-+-+-+
```

Figure 12.4. *AH extension format*

– Payload Length. This field, coded on 1 byte, provides the extension size in multiples of 4 bytes, not including the first 8 bytes. The size of the extension in IPv6 must remain a multiple of 8 bytes.

– SPI (Security Parameter Index). This field, coded on 4 bytes, contains a value pertaining to the previously negotiated security association.

– Sequence Number. This field, coded on 4 bytes, contains a value increased by one unit for each IPv4 or IPv6 packet transmitted. This field enables protection against replay. This field has a value of 1 for the first packet transmitted.

When the counter reaches the maximum value, a new security association must be negotiated in order to avoid the start of a new cycle.

An extended sequence number coded on 8 bytes constitutes an option making it possible for the lifetime of the security association to be prolonged. In order to preserve the structure of the extension, the 32 least significant bits are transmitted in the Sequence Number field. However, the data seal is calculated on all 64 bits.

– ICV (Integrity Check Value). This field is coded on a multiple of 4 bytes and contains the data seal ensuring authentication and integrity checking.

12.1.1.2. *ESP extension*

The ESP extension provides a confidentiality service in addition to those offered by the AH extension. The same ESP extension is used in association with an IPv4 or IPv6 header.

The presence of the extension is indicated by the Next Header (in IPv6) or Protocol (in IPv4) field of the previous header with a value of 50.

The ESP extension contains the same fields as the AH extension. It starts with the SPI and Sequence Number fields (Figure 12.5).

```
0                   1                   2                   3
0 1 2 3 4 5 6 7 8 9 0 1 2 3 4 5 6 7 8 9 0 1 2 3 4 5 6 7 8 9 0 1
+-+-+-+-+-+-+-+-+-+-+-+-+-+-+-+-+-+-+-+-+-+-+-+-+-+-+-+-+-+-+-+-+
|               Security Parameters Index (SPI)                 |
+-+-+-+-+-+-+-+-+-+-+-+-+-+-+-+-+-+-+-+-+-+-+-+-+-+-+-+-+-+-+-+-+
|                      Sequence Number                          |
+-+-+-+-+-+-+-+-+-+-+-+-+-+-+-+-+-+-+-+-+-+-+-+-+-+-+-+-+-+-+-+-+
|                    Payload Data (variable)                    |
~                                                               ~
|                                                               |
+               +-+-+-+-+-+-+-+-+-+-+-+-+-+-+-+-+-+-+-+-+-+-+-+-+
|               |     Padding (0-255 bytes)                     |
+-+-+-+-+-+-+-+-+               +-+-+-+-+-+-+-+-+-+-+-+-+-+-+-+-+
|                               |  Pad Length   | Next Header   |
+-+-+-+-+-+-+-+-+-+-+-+-+-+-+-+-+-+-+-+-+-+-+-+-+-+-+-+-+-+-+-+-+
|         Integrity Check Value-ICV   (variable)                |
~                                                               ~
|                                                               |
+-+-+-+-+-+-+-+-+-+-+-+-+-+-+-+-+-+-+-+-+-+-+-+-+-+-+-+-+-+-+-+-+
```

Figure 12.5. *ESP extension format*

After these fields, comes the encapsulated data, which may contain synchronization data Initialization Vector of the encryptor.

Following the encapsulated data, the extension ends with the following fields: Padding, Pad Length, Next Header and optionally ICV (authentication data).

The Padding field is necessary when block encryption is used and the block must be of a certain size, and to align the packet size with a multiple of 4 bytes.

12.1.1.3. *Modes*

For transport mode, the AH or ESP header is inserted between the IP header and the source IP packet payload. In the IPv6 environment, the AH or ESP header appears after the Hop-by-Hop, Destination, Routing and Fragment extensions.

For tunnel mode, the AH or ESP header encapsulates the source IP packet, and the whole is encapsulated in its turn by a new IP header. The tunnel corresponds to a data structure in which an IP packet contains another IP packet.

When the AH header is used, authentication is applied to the whole packet except the variable fields of the IP header, whose value is set to ZERO to calculate seal (Figure 12.6):

– DSCP (DiffServ Code Point): the value of this field can be modified by an intermediary router when it checks traffic characteristics;

– ECN (Explicit Congestion Notification): the value of this field can be modified by an intermediary router to alert the destination that congestion is developing;

– DF (Don't Fragment): this bit can be set at ONE by an intermediary router, when fragmented IP packet;

– Fragment Offset: insertion of the AH header occurs on non-fragmented IP packets, and therefore this field has a value of ZERO;

– TTL (Time to Live): the value of this field is decreased by one unit per each router crossed;

– Checksum: the value of this field is recalculated as soon as a field in the IP header changes value.

The group of IPv4 header options is considered as a single entity. Some options can be modified by an intermediary router. If a single modifiable option appears, the group of options is set at ZERO for the authentication digest calculation.

The variable fields of the IPv6 header are set at ZERO for the authentication digest calculation. These are identical fields to the ones in the IPv4 header (DSCP, ECN and Hop Limit), as well as the Flow Label field.

The Hop-by-Hop and Destination extensions of the IPv6 header have a bit that indicates whether the option can be modified by an intermediary router or not. If this bit is set at ONE, the extension is set at ZERO for the authentication digest calculation.

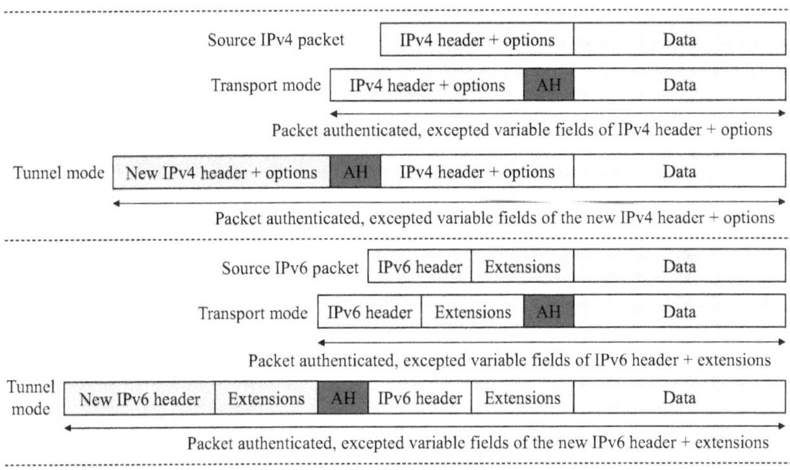

Figure 12.6. *Position of AH header*

When the ESP header is used in transport mode, the confidentiality service is applied to the encapsulated data and the ESP tail. Authentication and integrity services cover the ESP header, the encapsulated data and the ESP tail (Figure 12.7).

When the ESP header is used in tunnel mode, the confidentiality service is applied to the source IP packet and ESP tail. Authentication and integrity services cover the ESP header, the source IP packet and the ESP tail (Figure 12.7).

Note that in transport mode, the Destination extension can appear before, after or simultaneously before and after the AH or ESP extension.

12.1.2. *Security association*

A Security Association (SA) is a simple connection between two end points offering security services (confidentiality, integrity and authentication) to traffic.

Security services are provided by the use of AH or ESP extensions. To secure a two-directional communication between two end points, a security association pair is required. The Internet Key Exchange (IKE) protocol dynamically ensures the creation of the security association.

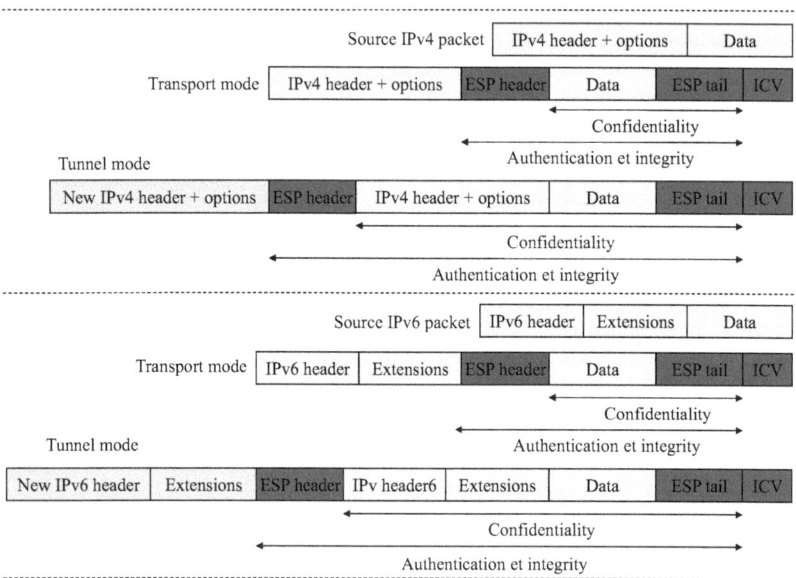

Figure 12.7. *Position of ESP header*

A security association contains the following parameters:

– the authentication algorithm and the key in order to generate the AH extension;

–the encryption algorithm and the key in order to generate the ESP extension;

– the authentication algorithm and the key in order to generate the ESP extension, if this service is used;

– the lifetime of the security association;

– the encapsulation mode (tunnel or transport).

IPSec mechanism defines three databases:

– SPD (Security Policy Database). This defines the security policy to be applied to input and output traffic for a host or a SEG;

– SAD (Security Association Database). This contains the parameters applied to an SA;

– PAD (Peer Authorization Database). This provides a link between the IKEv2 protocol and the SPD database.

The selector is the mechanism enabling identification at the source of the security association to be applied to traffic:

– Protocol or Next Header, source or destination IP address of the IP headers;

– source or destination port of the TCP (Transmission Control Protocol) or UDP (User Datagram Protocol) headers.

For outgoing packets, the selector consults the SPD database that defines the process to be applied to the packet:

– BYPASS: the packet is transmitted without a security service;

– DISCARD: the packet is deleted;

– PROTECT: the security service is applied to the packet. If the security association is not established, the IKE protocol is invoked. If the security association exists, the database returns a pointer to the SAD database.

The deletion of a packet causes the generation from the SEG toward the source of an ICMP message with the following characteristics:

– in IPv4 environment, Type = 3 (destination unreachable) and Code = 13 (Communication Administratively Prohibited);

– in IPv6 environment, Type = 1 (destination unreachable) and Code = 1 (Communication with destination administratively prohibited).

The SPI (Security Parameter Index) index is a field in the AH or ESP header used by the destination to identify the security association in a unique way. The destination uses this index to extract the security association from the SAD database.

The SPD database is consulted if the incoming packet is not protected, and the instruction (BYPASS or DISCARD) is applied.

12.1.3. *PMTU processing*

The introduction of ESP or AH extensions has an impact on the packet size, which can exceed the MTU (Maximum Transmission Unit) value. The PMTU (Path MTU) mechanism used to find the MTU value on the route makes it possible to avoid packet fragmentation by intermediary routers.

In IPv4 environment, the PMTU mechanism uses the DF bit in the IP header positioned at ONE. In the tunnel mode used between two SEGs, it is therefore necessary for the SEG to copy the value of this bit in the new header.

The PMTU mechanism uses an ICMP message indicating the MTU value. This message is sent back by an intermediary router, with the following characteristics:

– in IPv4 environment, Type = 3 (destination unreachable) and Code = 4 (Fragmentation needed and DF set);

– in IPv6 environment, Type = 2 (packet too big) and Code = 0 (Fragmentation needed).

The security gateway can thus receive this message from an unauthenticated source (an intermediary router) in the security association.

It must then analyze the content of the ICMP message to recover the information that will grant it access to the SAD database (IP addresses, SPI field).

It must send a new ICMP message to the host with a new PMTU value taking the size of the AH or ESP extensions into account.

12.2. IKEv2 protocol

IKEv2 protocol is more simplified than the previous version. It combines the functionalities defined in IKEv1 and ISAKMP (Internet Security Association and Key Management Protocol) while removing unnecessary processes.

It eliminates the generic character of the previous version, integrating the DOI (Domain of Interpretation) function, which defines the parameters specific to the ESP/AH security association.

Each IKEv2 message is composed of a header (HDR) and a sequence of blocks.The IKEv2 message is encapsulated by a UDP header with source and destination port values of 500 or 4,500. When the 4,500 port is used, the IKEv2 message is preceded by 4 bytes at ZERO.

12.2.1. *Message header*

The header of the IKE message contains the following fields (Figure 12.8):

```
                     1                   2                   3
 0 1 2 3 4 5 6 7 8 9 0 1 2 3 4 5 6 7 8 9 0 1 2 3 4 5 6 7 8 9 0 1
+-+-+-+-+-+-+-+-+-+-+-+-+-+-+-+-+-+-+-+-+-+-+-+-+-+-+-+-+-+-+-+-+
|                     IKE SA Initiator's SPI                    |
|                                                               |
+-+-+-+-+-+-+-+-+-+-+-+-+-+-+-+-+-+-+-+-+-+-+-+-+-+-+-+-+-+-+-+-+
|                     IKE SA Responder's SPI                    |
|                                                               |
+-+-+-+-+-+-+-+-+-+-+-+-+-+-+-+-+-+-+-+-+-+-+-+-+-+-+-+-+-+-+-+-+
| Next Payload  | MjVer | MnVer | Exchange Type |     Flags     |
+-+-+-+-+-+-+-+-+-+-+-+-+-+-+-+-+-+-+-+-+-+-+-+-+-+-+-+-+-+-+-+-+
|                          Message ID                           |
+-+-+-+-+-+-+-+-+-+-+-+-+-+-+-+-+-+-+-+-+-+-+-+-+-+-+-+-+-+-+-+-+
|                            Length                             |
+-+-+-+-+-+-+-+-+-+-+-+-+-+-+-+-+-+-+-+-+-+-+-+-+-+-+-+-+-+-+-+-+
```

Figure 12.8. *IKE message header format*

– Initiator's SPI: this field, coded on 8 bytes, incorporates a value chosen by the initiator. It initializes identification of the IKE security association.

– Responder's SPI: this field, coded on 8 bytes, incorporates a value chosen by the responder. It completes the identification of the IKE security association.

– Next Payload: this field, coded on 1 byte, incorporates the indication of the type of block following the header (Table 12.1).

– Major Version: this field, coded on 4 bits, indicates the maximum value of the IKE protocol version that can be used. This value is equal to 2 for thc implementation of the IKEv2 protocol.

– Minor Version: this field, coded on 4 bits, indicates the minimum value of the IKE protocol version. This value is equal to 0 for the implementation of the IKEv2 protocol.

Notation	Designation
SA	Security Association
KE	Key Exchange
Idi	Identification initiator
IDr	Identification responder
CERT	Certificate
CERTREQ	Certificate Request
AUTH	Authentication
Ni	Nonce initiator
Nr	Nonce responder
N	Notification
D	Delete
V	Vendor ID
TSi	Traffic Selector initiator
TSr	Traffic Selector responder
SK	Encrypted and Authenticated
CP	Configuration
EAP	Extensible Authentication

Table 12.1. *IKEv2 block types*

– Exchange Type: this field, coded on one byte, indicates the type of exchange to which the message belongs:

- IKE_SA_INIT: this exchange concerns the first phase of the establishment of the IKE security association;

- IKE_AUTH: this exchange concerns the second phase of the establishment of the IKE security association;

- CREATE_CHILD_SA: this exchange concerns the establishment of the ESP/AH security association;

- INFORMATIONAL: this exchange concerns event notification.

– Flags: This field includes the following three flags:

- R (Response): this flag, positioned at ONE, indicates that this message is a response. An IKE termination must not respond to a response except when authentication has failed;

- V (Version): this flag, positioned at ONE, indicates that the IKE termination is able to process a version higher than the one shown in the Major Version field;

- I (Initiator): this flag, positioned at ONE, indicates that the message is generated by the initiator of the IKE security association.

– Message ID: this field, coded on 4 bytes, is an identifier used to control the retransmission of lost messages and to correlate the request and the response. It also protects against replay attacks.

– Length: this field, coded on 4 bytes, includes the IKE message size.

12.2.2. *Blocks*

Each block starts with a generic header containing the Next Payload field, the C (Critical) bit and the Payload Length field (Figure 12.9).

```
 0                   1                   2                   3
 0 1 2 3 4 5 6 7 8 9 0 1 2 3 4 5 6 7 8 9 0 1 2 3 4 5 6 7 8 9 0 1
+-+-+-+-+-+-+-+-+-+-+-+-+-+-+-+-+-+-+-+-+-+-+-+-+-+-+-+-+-+-+-+-+
| Next Payload  |C|  Reserved   |         Payload Length          |
+-+-+-+-+-+-+-+-+-+-+-+-+-+-+-+-+-+-+-+-+-+-+-+-+-+-+-+-+-+-+-+-+
```

Figure 12.9. *Format of generic block header*

The Next Payload field indicates the type of block that comes next, thus enabling chaining.

The C bit determines the process to be executed when the receiver does not recognize the block:

– if the C bit is positioned at ONE, the receiver rejects the message;

– if the C bit is positioned at ZERO, the receiver ignores the block and processes the next block.

12.2.2.1. *SA block*

SA (Security Association) block is used for the negotiation of IKE and ESP/AH security association parameters.

SA block can contain several Proposals (P) ranked in order of preference. Each proposal defines a protocol (IKE, ESP or AH) and the value of the SPI index. Each proposal includes several Transformations (T) and each transformation includes one or more Attributes (A).

The transformation T involves the following operations:

– Encryption algorithm ENCR. This operation is used for negotiation pertaining to the IKE and ESP protocols.

– Pseudo-random function (PRF). This operation is used for negotiation pertaining to the IKE protocol.

– Integrity algorithm INTEG. This operation is used for negotiation pertaining to IKE, AH and optionally ESP protocols.

– Diffie–Hellman (D–H) group. This operation is used for negotiation pertaining to IKE and (optionally) AH and ESP protocols.

– Extended sequence number (ESN). This operation is used for negotiation pertaining to the AH and ESP protocols.

The attribute (A) specifies the length of the encryption algorithm key defined in the ENCR transformation. The other transformations, PRF, INTEG, D–H, and ESN, have no attributes.

12.2.2.2. *KE block*

The KE (Key Exchange) block contains the public Diffie-Helman value, enabling each end point (initiator and responder) to construct a shared secret. The block also mentions the DH-group defined in the SA block.

12.2.2.3. *IDi and IDr blocks*

The IDi (Identification initiator) and IDr (Identification responder) blocks contain an identification of the initiator of the IKE message and the responder. This identification is based on an IPv4 or IPv6 address, a name, a messaging address or a group of bytes.

12.2.2.4. *CERT block*

The CERT (Certificate) block provides a means of transporting a certificate or information pertaining to authentication.

12.2.2.5. *CERTREQ block*

The CERTREQ (Certificate Request) block is a request pertaining to a certificate. It is used in the response of the IKE_INIT_SA exchange response or in the IKE_AUTH exchange request. It also indicates the certification authority for the required certificate.

12.2.2.6. *AUTH block*

The AUTH (Authentication) block contains the authentication digest of the initiator or the responder. The block also specifies the method used.

12.2.2.7. *Ni and Nr blocks*

The Ni (Nonce initiator) and Nr (Nonce responder) blocks contain a random number generated by the initiator and the responder. These numbers are used in the creation of derived keys.

12.2.2.8. *N block*

The N (Notification) blocks contain error messages indicating the reason why the SA cannot be established.

The N block also contains status messages that an SA management process wishes to communicate to a remote process.

12.2.2.9. *D block*

The D (Delete) block includes the SPI index of the SA that the message source wishes to delete. For an AH or ESP protocol, it is possible to specify several index values. It is also possible to string several D blocks together in a single IKEv2 message.

12.2.2.10. *V block*

The V (Vendor ID) block announces that the message source is capable of accepting private extensions of the IKEv2 protocol. These extensions can involve the introduction of new blocks, new types of exchange or new notification information.

12.2.2.11. *TS block*

The TS (Traffic Selector) block identifies the flows for which the ESP/AH security association is implemented. Flow determination is based on the following information:

– the type of data encapsulated by the IP header, stated in the Protocol field of the IPv4 header or the Next Header field of the IPv6 header;

– the range of source and destination IP addresses;

– the range of source and destination port numbers if the IP header encapsulates UDP or TCP segments;

– the ICMP message type and code.

12.2.2.12. *SK block*

The SK (Encrypted and Authenticated) block is always located at the end of the IKEv2 message. The encryption and integrity algorithms of the IKEv2 algorithms are negotiated during the implementation of the IKEv2 security association.

12.2.2.13. *CP block*

The CP (Configuration) block is used to exchange configuration information between the two end points. In the case of an ESP/AH security association between a host and a SEG, the host can request information concerning a host in the protected network.

12.2.2.14. *EAP block*

The EAP (Extensible Authentication Protocol) block enables authentication of the IKE security association by the EAP protocol.

12.2.3. *Procedure*

12.2.3.1. *IKE_SA_INIT exchange*

The first exchange, IKE_SA_INIT, negotiates cryptographic algorithms and random numbers and executes a Diffie-Hellman exchange in order to create an IKE security association.

The initiator generates an IKE message containing the header (HDR) and the SAi1, KEi and Ni blocks. The header (HDR) contains the SPI index of the initiator, the version numbers and the flags. The SAi1 block contains the cryptographic algorithms proposed by the initiator for the IKE security association. The KEi block includes the Diffie-Hellman group and the public value. The Ni block displays the initiator's random number (Figure 12.10).

The responder chooses a cryptograph series from the initiator's proposals and includes it in the SAr1 block. It completes the exchange of Diffie-Hellman keys with the KEr block. It sends its random number in the Nr block (Figure 12.10). It can possibly communicate a list of certificate authorities in the CERTREQ block.

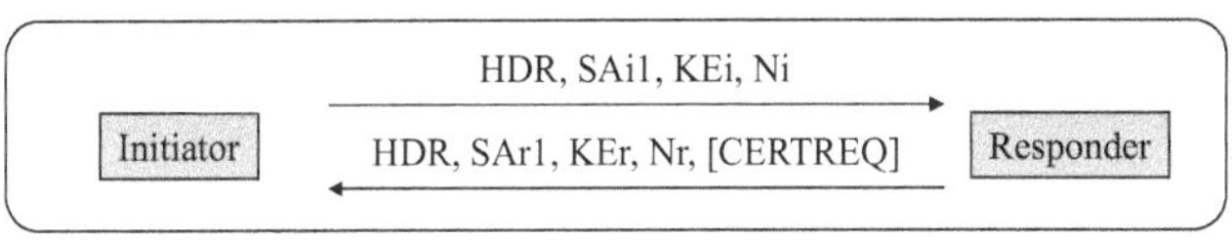

Figure 12.10. *IKE_SA_INIT exchange*

At this stage of the negotiation, each end point can generate the SKEYSEED key. The keys used for encryption and integrity of IKE messages are produced by the SKEYSEED key and are known as SK_e (encryption) and SK_a (integrity). Message protection involves only the blocks; the header is not included.

The two different directions of traffic use different keys. The keys used to protect messages from the initiator are SK_ai and SK_ei.

The keys used to protect messages in the other direction are SK_ar and SK_er.

Other keys are also derived from the SKEYSEED key. The SK_d key is used to derive the keys used in the ESP/AH security association. The SKEYSEED key is calculated using the Diffie-Hellman secret (D–H key) and the random numbers Ni and Nr.

SKEYSEED = PRF (D–H key, Ni | Nr)

The keys SK_d, SK_ai, SK_ar, SK_ei, SK_er, SK_pi and SK_pr are generated as follows:

SK_d = PRF (SKEYSEED, Ni | Nr | SPIi | SPIr | 0x01)

SK_ai = PRF (SKEYSEED, SK_d | Ni | Nr | SPIi | SPIr | 0x02)

SK_ar = PRF (SKEYSEED, SK_ai | Ni | Nr | SPIi | SPIr | 0x03)

SK_ei = PRF (SKEYSEED, SK_ar | Ni | Nr | SPIi | SPIr | 0x04)

SK_er = PRF (SKEYSEED, SK_ei | Ni | Nr | SPIi | SPIr | 0x05)

12.2.3.2. *IKE_AUTH exchange*

The second exchange, IKE_AUTH, is used to authenticate previous IKE messages and communicate identities and possibly exchange certificates, as well as to establish the first AH/ESP security association. These messages are completely encrypted and protected by the keys established during the IKE_SA_INIT exchange.

The initiator indicates its identity with the IDi block. It authenticates its identity and protects the integrity of the first message in the IKE_SA_INIT exchange using the AUTH block.

It can possibly send its certificate in the CERT block and the identity of the certification authority in the useful CERTREQ payload load. In this case, AUTH contains the signature of the initiator (Figure 12.11).

The optional IDr block enables the initiator to specify which of the identities of the responder it wishes to communicate with (Figure 12.11).

The initiator starts the AH/ESP security association negotiation with the SAi2 block. The TSi block specifies the characteristics of the packets transferred by the initiator. The TSr block specifies the characteristics of the packets transferred by the responder (Figure 12.11).

The notation SK{ ... } indicates that the blocks are completely encrypted and protected (Figure 12.11).

The responder communicates its identity in the IDr block. It may send a certificate. It authenticates its identity and protects the integrity of the second message of the IKE_SA_INIT exchange. It completes the negotiation of the ESP/AH security association with the SAr2 block (Figure 12.11).

Figure 12.11. *IKE_AUTH exchange*

12.2.3.3. *CREATE_CHILD_SA exchange*

The CREATE_CHILD_SA exchange is used to create the ESP/AH security association and to renew IKE and ESP/AH security association keys.

For the exchange involving the creation of the ESP/AH security association, the initiator finalizes the security association in the SA block and the traffic selectors proposed for the SA in the TSi and TSr blocks. It transmits a random number in the Ni block and optionally a Diffie-Hellman value in the KEi block (Figure 12.12).

The responder confirms the security association offer in the SA block. It transmits a Diffie-Hellman value in the KEr block, if the KEi block has been included in the request (Figure 12.12).

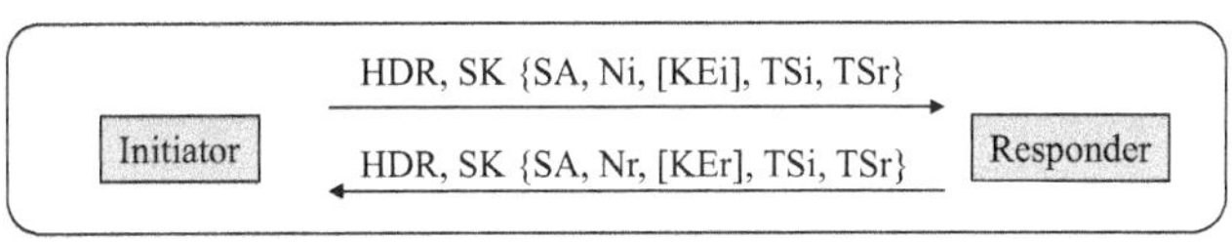

Figure 12.12. *CREATE_CHILD_SA exchange*
Creation of ESP /AH security association

The KEYMAT key, used for the ESP/AH security association, is derived from the SK_d key and the random numbers Ni and Nr. If the exchange contains Diffie-Hellman values, the secret Diffie-Hellman key obtained also participates in the creation of the KEYMAT key.

KEYMAT = PRF (SK_d, Ni | Nr)

KEYMAT = PRF (SK_d, D-H key | Ni | Nr)

To renew the IKE security association, the initiator sends the security association in the SA block, a random number in the Ni block and a Diffie-Hellman value in the KEi block. The new SPI index of the initiator is provided in the SA block (Figure 12.13).

The responder confirms the offer in the SA block. It transmits a Diffie-Hellman value in the KEr block. The new SPI index of the responder is provided in the SA block (Figure 12.13).

The new SKEYSEED key is calculated from the old SK_d key, the D-H secret key and the random numbers Ni and Nr.

SKEYSEED = PRF (previous SK_d, D-H key | Ni | Nr)

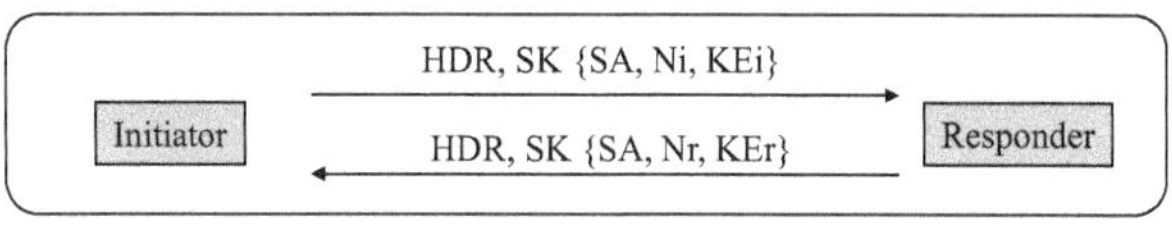

Figure 12.13. *CREATE_CHILD_SA exchange Renewal of the ESP/AH security association*

To renew the ESP/AH security association key, the messages transmitted by the initiator and the responder are similar to the ones used in the creation of the security association.

The request of the initiator contains an N block (REKEY_SA) containing the SPI index value of the new security association (Figure 12.14).

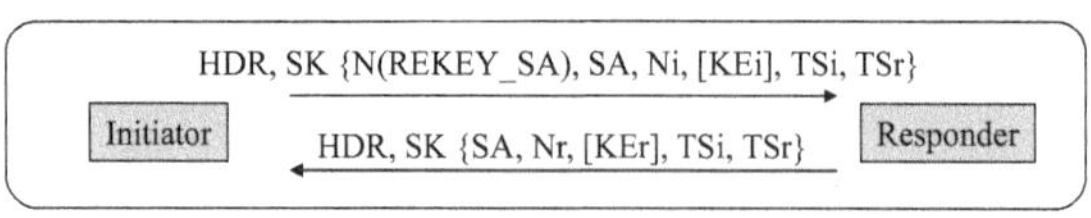

Figure 12.14. *CREATE_CHILD_SA exchange Renewal of the key of the ESP/AH security association*

12.4. Security architecture over S1 interface

The enrolment process of the eNB entity to the RA/CA (Registration Authority/Certification Authority) entity is to dynamically establish the certificate of the eNB entity certified by the operator.

The eNB entity is initially configured by the manufacturer with a pair of public/private keys. The public key of the eNB entity is included in a certificate signed with the private key of the manufacturer (Figure 12.15).

The RA/CA entity is provisioned with the root certificate of the manufacturer. This certificate contains the public key of the manufacturer and it is signed with the private key of the manufacturer (Figure 12.15).

The SEG entity is configured by the operator with a public key and a private key. The public key of the SEG entity is included in a certificate signed with private key of the operator.

The SEG entity is provisioned with the root certificate of the operator. This certificate contains the public key of the operator and it is signed with the private key of the operator (Figure 12.15).

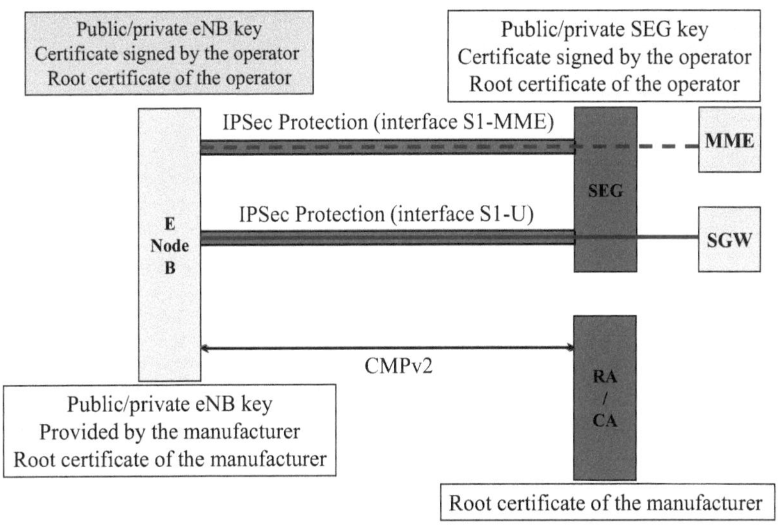

Figure 12.15. *Security architecture over S1 interface. For a color version of this figure, see www.iste.co.uk/perez/4gmobile.zip*

After the enrolment process, the eNB entity is provisioned with the following data (Figure 12.15):

– the new pair of public/private keys to be used for IPSec mechanism;

– the certificate containing the new public key, signed with the private key of the operator;

– the root certificate of the operator containing the public key of the operator.

12.4.1. *RA/CA entity*

The RA entity checks that the eNB entity has the right to register, authenticates CMPv2 (Certificate Management Protocol) messages exchanged with the eNB entity and sends the certificate request to the CA entity.

The RA entity authenticates the eNB entity from its signature and its certificate signed by the manufacturer. The signature is performed using the private key of the eNB entity installed by the manufacturer. The certificate signed by the manufacturer contains the public key of the eNB entity installed by the manufacturer.

The RA entity can attribute to the entity eNB an identity defined by the operator.

The CA entity generates the certificate of the eNB entity signed by the operator, and containing the public key generated by the eNB entity.

12.4.2. *Enrolment procedure*

The enrolment procedure of the eNB entity to the RA/CA entity is depicted in Figure 12.16.

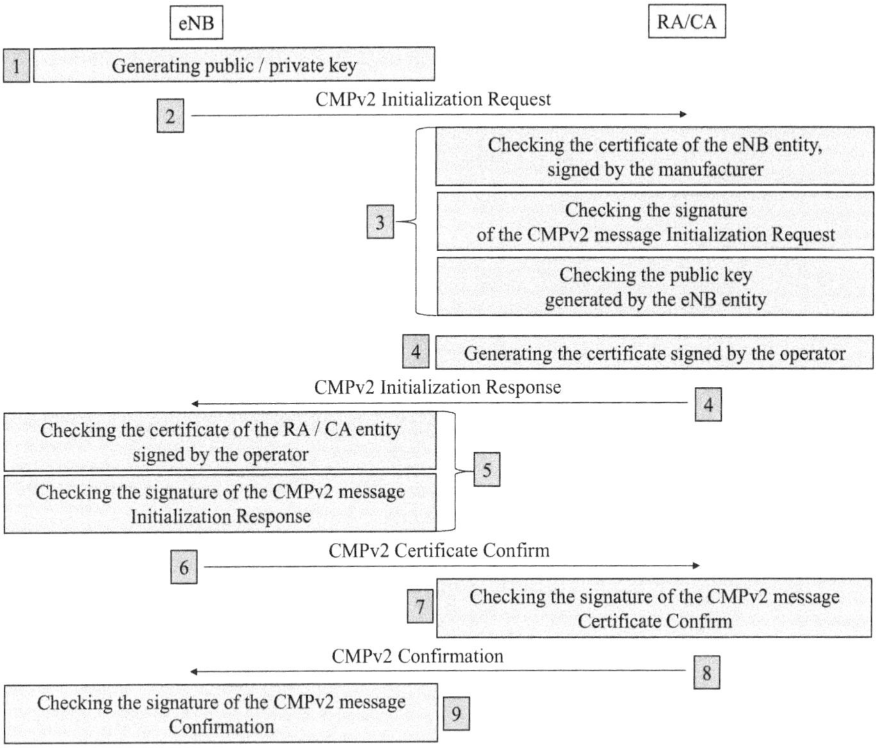

Figure 12.16. *Enrolment procedure of the eNB entity*

1) The eNB entity generates the new private/public key pair to be used for the IPSec mechanism.

2) The eNB entity transmits to the RA/CA entity the CMPv2 message Initialization Request containing the public key generated by the eNB entity, signed with the private key generated by the eNB entity to prove it is the holder of the public key.

The CMPv2 message Initialization Request is signed with the private key that the manufacturer has provided to the eNB entity.

The CMPv2 message Initialization Request contains the certificate of the eNB entity signed by the manufacturer containing the public key provided by the manufacturer.

3) The RA/CA entity verifies the certificate signed by the manufacturer with the public key contained in the root certificate of the manufacturer.

The RA/CA entity authenticates the CMPv2 message Initialization Request with the public key contained in the certificate of the eNB entity signed by the manufacturer.

The RA/CA entity checks the signature of the public key generated by the eNB entity contained in the CMPv2message Initialization Request.

4) The RA/CA entity generates a certificate signed by the operator containing the public key generated by the eNB entity. This certificate is transmitted to the eNB entity in the CMPv2message Initialization Response.

The CMPv2 message Initialization Response is signed with the private key of the RA/CA entity.

The CMPv2 message Initialization Response also contains the certificate of the RA/CA entity signed by the operator and the root certificate of the operator.

5) The eNB entity checks the certificate of the RA/CA entity with the public key contained in the root certificate of the operator.

The eNB entity authenticates the CMPv2 message Initialization Response with the public key of the RA/CA entity contained in its certificate.

6) The eNB entity confirms the receipt of the certificate in the CMPv2 message Confirm Certificate. This message is signed with the private key generated by the eNB entity.

7) The RA/CA entity authenticates the CMPv2 message Certificate Confirm with the public key generated by the eNB entity.

8) The RA/CA entity confirms the receipt of the CMPv2 message Certificate Confirm by sending the CMPv2 message Confirmation signed with the private key of the RA/CA entity.

9) The eNB entity authenticates the CMPv2message Confirmation with the public key of the RA/CA entity.

13

Synchronization Principles

13.1. Synchronization types

Frequency synchronization is obtained by the distribution of the reference rhythm to all nodes of the network (Figure 13.1).

Phase synchronization means that the network nodes acquire a reference signal from which the significant instants (typically the beginning of a time frame) occur simultaneously. The phase synchronization permits frequency synchronization (Figure 13.1).

Time synchronization is the time distribution to all nodes of the network, which corresponds to a time stamp of the significant instants. Time synchronization permits for phase synchronization and frequency (Figure 13.1).

The method based on synchronous physical layer is a method suitable for frequency synchronization. The rhythm information is contained in the physical signal and does not depend on the traffic load.

The method of synchronous physical layer is a point-to-point method. Each intermediate node must be part of the rhythm signal distribution system.

The packet-based method used in bidirectional mode allows a time, phase and frequency synchronization. Traffic has an impact on performance, especially in the absence of a behavior management mechanism based on classes of service.

For the packet-based method, nodes not involved in the time distribution can be crossed.

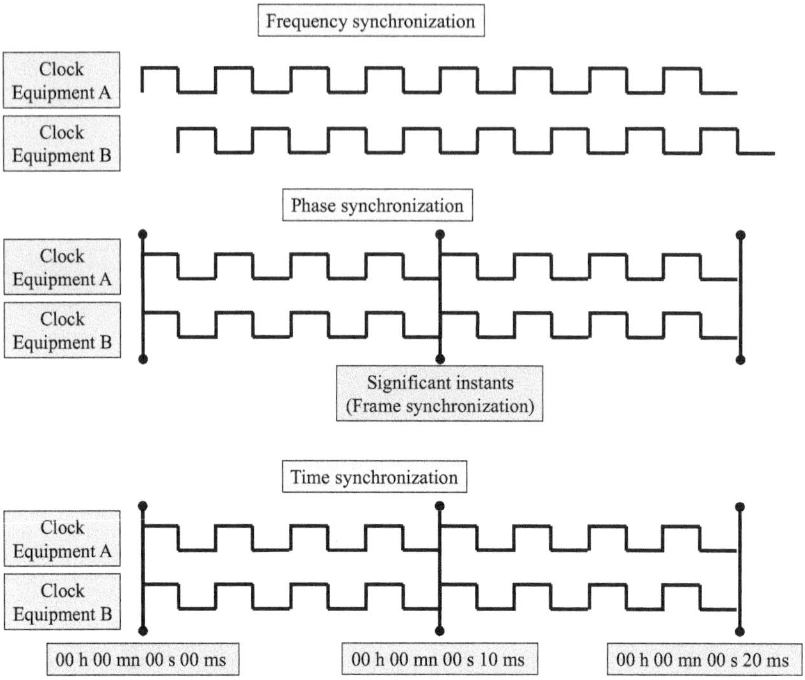

Figure 13.1. *Synchronization types*

SyncE (Synchronous Ethernet) mechanism, described in Chapter 14, uses the method based on synchronous physical layer.

IEEE 1588 mechanism, described in Chapter 15, uses the packet-based method.

13.2. Clock characteristics

13.2.1. *Method based on synchronous physical layer*

The reference rhythm is generated by a reference oscillator and then distributed by the transport network. A Phase-Locked Loop (PLL) allows for the recovery of the reference rhythm from the physical Ethernet signal at each equipment.

The phase noise, produced by the oscillators and by the propagation conditions of the physical Ethernet signal, characterizes the quality of the reference signal distributed by the transport network.

Phase noise is represented in the time domain by variation of the position of the front of the clock with respect to a reference clock (Figure 13.2).

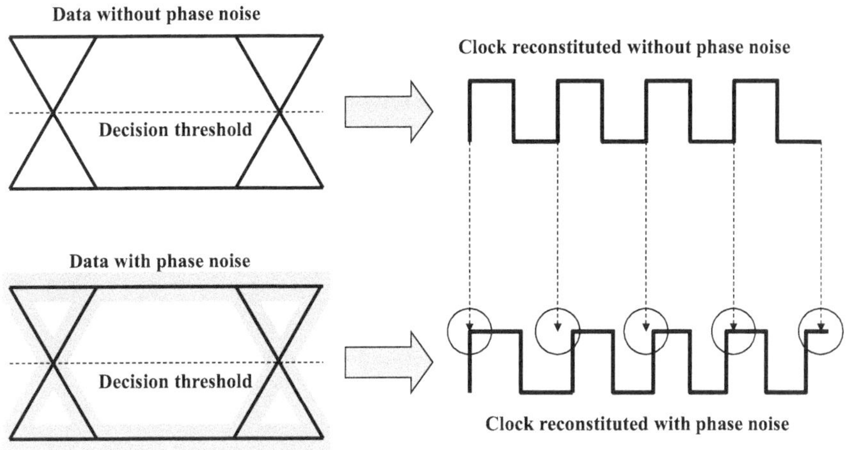

Figure 13.2. *Representation of phase noise in the time domain*

Phase noise is represented in the frequency domain by a widening of the spectrum of the clock relative to an ideal frequency characterized by a ray.

Jitter is represented by short-term variations of the significant instants of the clock compared to the same nominal values, the meaning of short term corresponding to variations at a frequency greater than or equal to 10 Hz (Figure 13.3).

The jitter is measured in maximum phase amplitude. The unit used is the time interval. A time interval corresponds to the duration of the bit of the physical signal. The jitter measurement UIpp (Unit Interval peak-to-peak) corresponds to a peak-to-peak value of the amplitude.

The wander is represented by the long-term variations of the significant instants of the clock compared to the same nominal values, the meaning of long-term corresponding to variations at a frequency less than 10 Hz (Figure 13.3).

Temperature changes, firstly, between day and night and, secondly, between seasons are the main causes of the wander:

– they cause a variation in the speed of propagation of the electrical or optical signal;

– they create a variation in the length of the physical medium;

– they cause different behavior for each oscillator.

The measurement of the wander of a clock produced by a synchronous physical layer is quantified by the following parameters: Maximum Time Interval Error (MTIE) and Time Deviation (TDEV) (Figure 13.3).

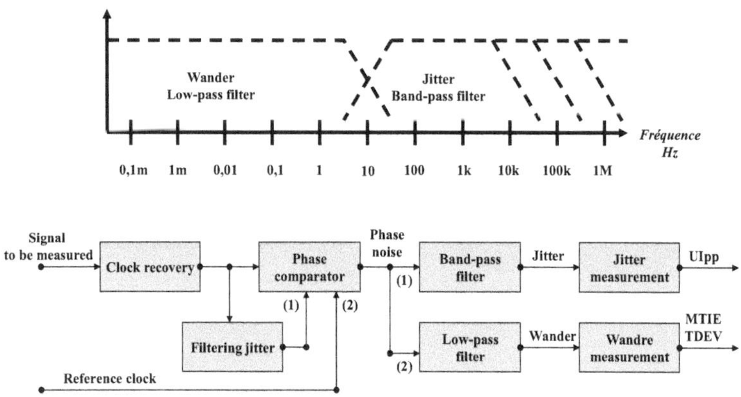

Figure 13.3. *Phase noise measurements*

The time error x(t) of a clock, with respect to a frequency standard, is the difference between the time of that clock T(t) and the frequency standard one $T_{ref}(t)$.

$$x(t) = T(t) - T_{ref}(t)$$

Time Interval Error (TIE) is the difference between the measure of a time interval (τ) as provided by a clock and the measure of that same time interval as provided by a reference clock.

$$TIE(t,\tau) = [T(t+\tau) - T(t)] - [T_{ref}(t+\tau) - T_{ref}(t)] = x(t+\tau) - x(t)$$

where τ represents the time interval, called the observation period.

Maximum Time Interval Error (MTIE) is the maximum peak-to-peak delay variation of a given timing signal with respect to an ideal timing signal within an observation time ($\tau = n\tau_0$) for all observation times of that length within the measurement period (T).

$$\mathrm{MTIE}(n\tau_0) = \max_{1\le k\le N-n}\left[\max_{k\le i\le k+n} x_i - \min_{k\le i\le k+n} x_i\right], \text{ with } n = 1,2,\ldots.N-1$$

where:

- x_i is the ith time error sample;
- N is the total number of samples;
- τ_0 is the sample period;
- n is the number of sampling intervals;
- $\tau = n\tau_0$ is the observation time.

Time Deviation (TDEV) is a measure of the expected time variation of a signal as a function of integration time. TDEV can also provide information about the spectral content of the phase noise of a signal.

$$\mathrm{TDEV}(n\tau_0) \cong \sqrt{\frac{1}{6n^2(N-3n+1)}\sum_{j=1}^{N-3n+1}\left[\sum_{i=j}^{n+j-1}(x_{i+2n} - 2x_{i+n} + x_i\right]^2}$$

with $n = 1,2,\ldots.$integer part $\left(\frac{N}{3}\right)$

13.2.2. *Packet-based method*

The time reference is generated by a reference clock, and then transmitted in packets carried by the transport network.

The damage in the transport network that affects the performance of a packet-based clock is the packet loss and the Packet Delay Variation (PDV) related to the physical layer and packet buffering.

13.3. Synchronization requirements

The different entities of the mobile network, except the evolved Node Base station (eNB), and the transport network require no synchronization to implement data transfer.

The radio interfaces of adjacent eNB entities must be synchronized in frequency to ensure the change in cells during a session (handover) for both modes of operation: the Frequency Division Duplex (FDD) and Time Division Duplex (TDD).

For FDD mode, transmission and reception are performed in a frequency band allocated to each direction of transmission. The transmission and reception can, therefore, operate simultaneously.

For the TDD mode, transmission and reception are performed in the same frequency band. The resource is shared in time for each direction of transmission. The transmission and reception cannot, therefore, operate simultaneously.

For TDD mode, phase synchronization is necessary to synchronize the times of transmission and reception of adjacent eNB entities.

Interference management between eNB entities, based on a time sharing of the radio resource, also requires phase synchronization of adjacent eNB entities.

13.3.1. *Operation modes*

13.3.1.1. *FDD mode*

The type 1 structure defined for the FDD mode has duration of 10 ms and contains 10 subframes (Figure 13.4). Each subframe consists of two time slots.

The frequency synchronization of eNB entities is required for the FDD mode to ensure the change in cells during a session (handover). A frequency offset between the source and target eNB entities results in a delay or a failure of locking the mobile on the frequency of the adjacent eNB entity. These dysfunctions may impact the handover procedure.

The same source of synchronization is used for the generation of the subcarriers and the bit rate on the radio interface.

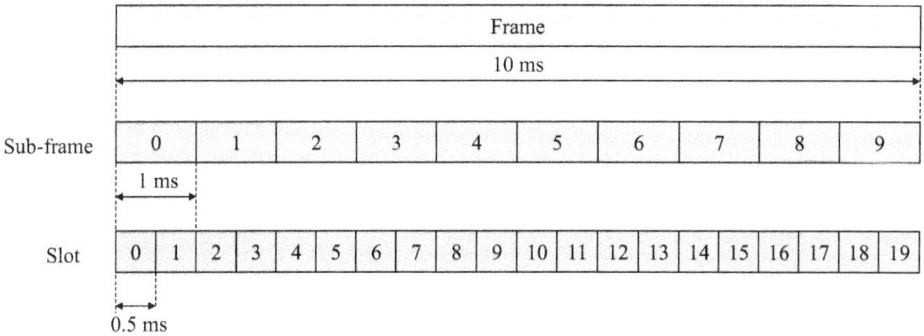

Figure 13.4. *Structure of the frame in FDD mode*

The accuracy of the clock required on the radio interface is a function of the cell size (Table 13.1).

Cell type	Clock accuracy (ppm)
Macro-cell	±0.05 ppm
Micro-cell, Pico-cell	±0.1 ppm
Femto-cell	±0.25 ppm

Table 13.1. *Accuracy of the clock required for the handover*

The jitter limit in the network at the input interface of the eNB entity is less than 1.5 UIpp for Ethernet interfaces to 1 Gbps or 10 Gbps.

The wander limit in the network at the input interface of the eNB entity, expressed as MTIE, is given in Table 13.2.

Observation interval τ (s)	MTIE requirement (ns)
$0.1 < \tau \leq 2.5$	250
$2.5 < \tau \leq 20$	100τ
$20 < \tau \leq 2{,}000$	2,000
$\tau > 2{,}000$	$433\tau^{0.2} + 0.01\tau$

Table 13.2. *MTIE requirement at the input interface of the eNB entity*

The wander limit in the network at the input interface of the eNB entity, expressed as TDEV, is given in Table 13.3.

Observation interval τ (s)	TDEV requirement (ns)
$0.1 < \tau \leq 17.14$	12
$17.14 < \tau \leq 100$	0.7τ
$100 < \tau \leq 1{,}000{,}000$	$58 + 1.2\tau^{0.5} + 0.0003\tau$

Table 13.3. *TDEV requirement at the input interface of the eNB entity*

13.3.1.2. *TDD mode*

The type-2 structure defined for the TDD mode also has duration of 10 ms and contains two half-frames of 5 ms each (Figure 13.5).

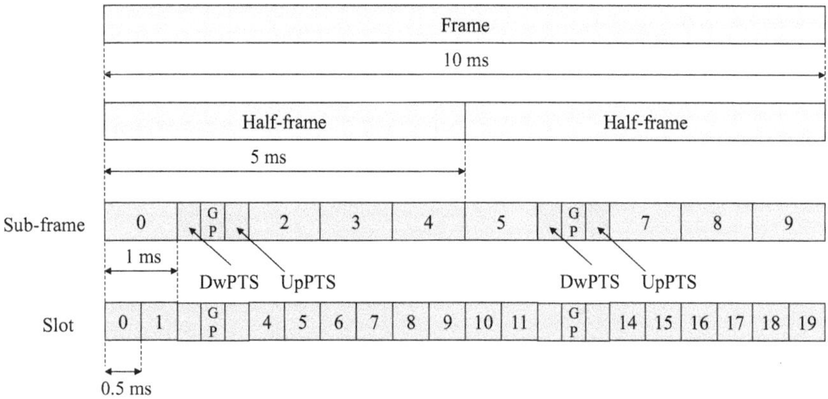

Figure 13.5. *Structure of the frame in TDD mode*

Each half-frame comprises five subframes and the second subframe can correspond to a special subframe containing three particular fields:

– a field for the Downlink Pilot Time Slot (DwPTS). This field can contain data.

– a field for the Uplink Pilot Time Slot (UpPTS). This field can contain data or a preamble.

– a Gap Period (GP) between the two previous fields. This interval time facilitates the compensation of a time difference between different mobiles and avoids an overlap between the two transmission directions.

The subframes are attributed to the data for the uplink and downlink according to diverse configurations (Table 13.4)

Configuration	Periodicity (ms)	Number of the subframe									
		0	1	2	3	4	5	6	7	8	9
0	5	D	S	U	U	U	D	S	U	U	U
1	5	D	S	U	U	D	D	S	U	U	D
2	5	D	S	U	D	D	D	S	U	D	D
3	10	D	S	U	U	U	D	D	D	D	D
4	10	D	S	U	U	D	D	D	D	D	D
5	10	D	S	U	D	D	D	D	D	D	D
6	5	D	S	U	U	U	D	S	U	U	D

D, downlink; U, uplink; S, special.

Table 13.4. *Configuration of the TDD frame*

Sub-frames 0 and 5 are always allocated to traffic for the downlink.

Sub-frame 1 is always allocated to the special sub-frame containing the three particular fields.

Sub-frame 2 is always allocated to traffic for the uplink.

Sub-frame 6 can be allocated to the special sub-frame containing three particular fields for a periodicity of 5 ms.

Sub-frames 3, 4, 7, 8, 9 are allocated to the downlink or uplink traffic according to the selected configuration.

13.3.1.3. *Time slot structure*

Each time slot comprises 3, 6 or 7 OFDM (Orthogonal Frequency Division Multiplexing) symbols (Figure 13.6).

The introduction of a guard time between the symbols facilitates the removal of intersymbol interference. The guard time of a time slot is found at the beginning of the symbol.

It contains the copy of the end of the symbol in order to avoid a very high dynamic for the amplifiers. This copy is called Cyclic Prefix (CP) (Table 13.5).

The normal cyclic prefix is used in case the delay between the different reflected signals is low, which is the case in cells of small diameter (micro-cell, pico-cell, femto-cell).

The extended cyclic prefix is used in case the delay between the different reflected signals is significant, which is the case in cells of a large diameter (macro-cell).The frequency synchronization of the eNB entities is also required for the TDD mode to ensure the change in cells during a session (handover).

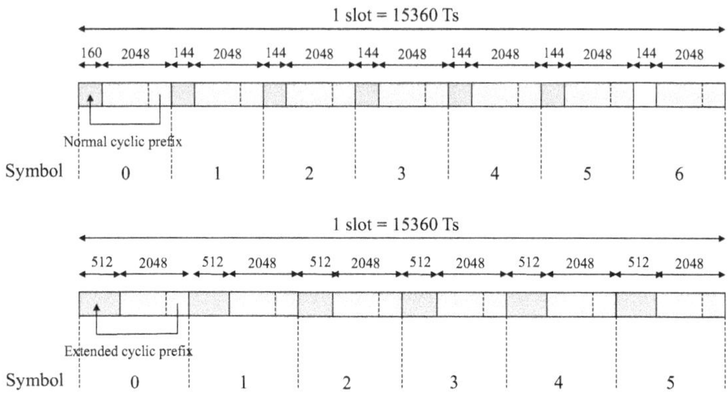

$T_s = 1/(15{,}000 \times 2{,}048)$ s

Figure 13.6. *Time slot structure*

Downlink		Length of the cyclic prefix
Normal cyclic prefix	Δf = 15 kHz	5.2 µs for l = 0 4.68 µs for l = 1,2,...,6
Extended cyclic prefix	Δf = 15 kHz	16.66 µs for l = 0,1,...,5
	Δf = 7.5 kHz	33.33 µs µs for l = 0,1,2
Uplink		Length of the cyclic prefix
Normal cyclic prefix		5.2 µs for l = 0 4.68 µs for l = 1,2,...,6
Extended cyclic prefix		16.66 µs for l = 0,1,...,5

Δf, step between subcarriers; l, number of the symbol in the time slot.

Table 13.5. *Length of the cyclic prefix*

The phase synchronization of the eNB entities results in a synchronous transmission of the eNB entities, at the frame level, in order to align the periods of transmission and reception.

The tolerance required on the alignment of the frame depends on the length of the cyclic prefix (Table 13.6).

Cyclic prefix type	Tolerance of alignment (μs)
Normal cyclic prefix	≤3
Extended cyclic prefix	≤10

Table 13.6. *Tolerance of alignment*

13.3.2. *Interference management*

A homogeneous network is a network for which an area is covered by adjacent cells, generally with the same size. The interference area between the cells is limited to the edge of the cell (Figure 13.7).

A heterogeneous network is a network for which an area is covered simultaneously by cells of different sizes (macro-cell, micro-cell, pico-cell, femto-cell). For example, the interference zone is constituted by the pico-cell cell located within a macro-cell cell (Figure 13.7).

Coordinated Multi-Point operation (CoMP) can improve cell performance when the mobile is in an interference area between cells that operate in the same frequency band.

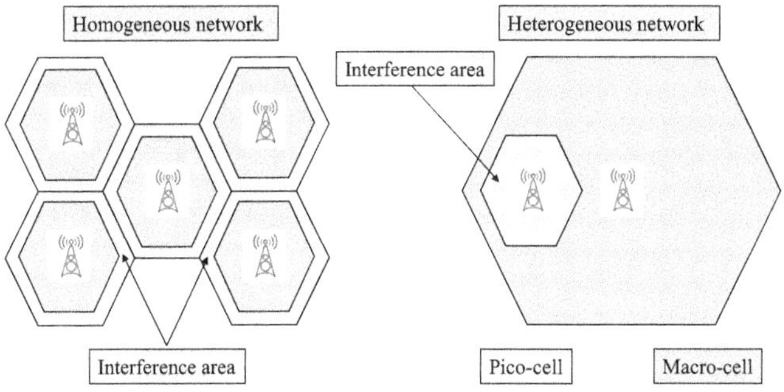

Figure 13.7. *Interference areas*

For the downlink direction, three mechanisms according to the CoMP function are defined:

– JT (Joint Transmission) mechanism is based on the simultaneous transmission of the same signal by different eNB entities, which allows to remove the interference and improve the quality of the received signal.

– DPS (Dynamic Point Selection) mechanism is based on the transmission of the signal by a single eNB entity, although the same signal is transmitted to eNB several entities.

– CS/CB (Coordinated Scheduling/Beamforming) mechanism is based on the transmission of the signal by a single eNB entity, the signal not being transmitted to several eNB entities. Coordination between the eNB entities of the frequency band allocation and precoding for beamforming toward the mobile used to mask interference and improve cell performance.

For the uplink direction, two mechanisms according to the CoMP function are defined:

– JR (Joint Reception) mechanism is based on the reception by several eNB of the signal transmitted by the mobile, which improves the quality of the received signal.

– CS/CB mechanism determines, for a set of eNB entities, the frequency band allocated and precoding for beamforming. On the other hand, the signal transmitted by the mobile is processed by a single eNB entity.

JT and DPS mechanisms for the downlink direction and JR mechanism for the uplink direction require phase synchronization of the eNB entities involved in the CoMP function.

14

Synchronization
SyncE Mechanism

14.1. Synchronization network architecture

14.1.1. *Entities*

SyncE (Synchronous Ethernet) mechanism is an extension of the architecture of reference clock defined for SDH (Synchronous Digital Hierarchy) or SONET (Synchronous Optical Network).

The reference clock traceable to a Primary Reference Clock (PRC) is injected into the equipment having physical Ethernet interfaces by an external interface T3 at 2048 kHz or 2048 kbps.

The PRC clock determines the long-term stability of the reference frequency.

The PRC clock is generated from a cesium-beam tube that provides a maximum frequency shift relative equal to 10^{-11} for observation periods of more than a week.

The synchronization network may run two (or more) PRC at different locations in order to achieve a very high availability of reference signals to the network. In case of a fault of the currently active PRC, the standby PRC will take over the role of the failed PRC.

Radio-controlled PRCs use remote cesium tubes, e.g. in the satellites of the GPS navigation system. In this case, radio signals are generated on the basis of cesium tubes received at the location of the PRC.

The synchronization function, supported by Ethernet Equipment Clock (EEC), provides total filtering of jitter and partial filtering of wander.

Filtering is provided by a Phase-Locked Loop that constitutes a low-pass filter whose cutoff frequency is equal to 5 Hz (Figure 14.1).

The reference clock T0 of the EEC function is generated by the synchronization function before being used for the Ethernet interfaces TE or T4 (output synchronization interface).

The distribution of the reference clock may involve a number of equipment in cascade. In this case, the EEC function must retrieve the reference clock from the input Ethernet interface TE.

The clock of EEC function is generally synchronized with the T3 or TE interface (switch B of Figure 14.1).

In the case of a stand-alone operation, the clock of the EEC function provides initial accuracy better than 4.6×10^{-6} and a maximum relative frequency shift equal to 10^{-8} for observation periods of more than one day. The clock has a moderate frequency accuracy in holdover mode in order to allow for moderate cost oscillator.

EEC function provides a downstream reference clock at T4 interface. This signal is based on Ethernet interfaces TE or on the reference clock T0 (switches A and C of Figure 14.1).

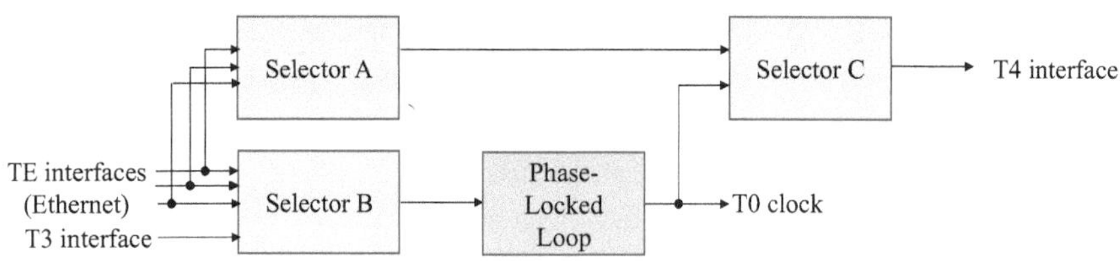

Figure 14.1. *EEC function*

Synchronization Supply Unit (SSU) is a clock incorporated in the chain of EEC functions to ensure partial filtering the wander (low-pass filter whose cutoff frequency is equal to 3 mHz).

An SSU is normally not connected directly to traffic signals. The node terminating the transport of the traffic signals extract a reference clock out and deliver it on the T4 interface.

In the case of a stand-alone operation, the SSU clock provides an initial accuracy better than 5×10^{-10} and a maximum relative frequency shift equal to 2×10^{-10} for observation periods more than one day.

The SSU function provides a reference clock to the downstream EEC functions on T3 interface.

14.1.2. *Synchronization chain*

The reference chain of the synchronization network determined, for the worst case, the following maximum values (Figure 14.2):

– 60 EEC clocks;

– 20 clocks EEC between 2 SSU clocks;

– 10 SSU clocks.

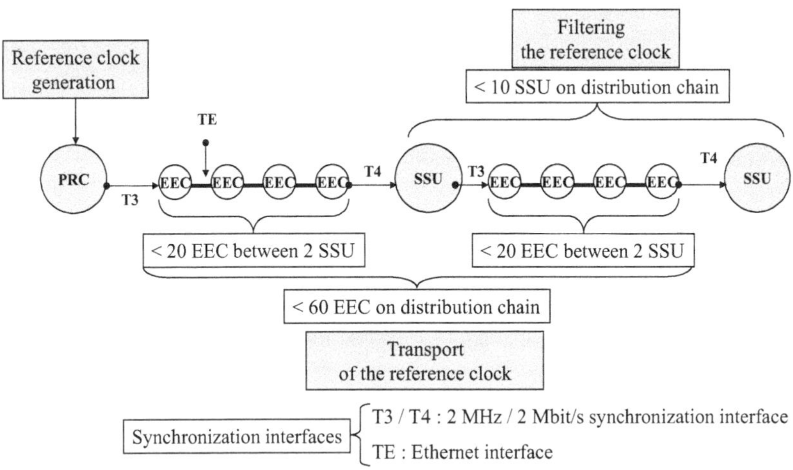

Figure 14.2. *Transport the reference clock and filtering*

In the case where the Ethernet signal TE is carried on the Optical Transport Network (OTN), it is considered that the OTN and EEC functions are not synchronized. In this case, the GFP-F (Generic Framing Procedure-Frame mapped) protocol allows the mapping of the Ethernet frame in an OPU (Optical channel Payload Unit) frame.

The GFP-T (Generic Framing Procedure - Transparent mapped), protocol, which transports the clock signal, can be used to map the Ethernet signal in an OPU frame, if the EEC functions and the OTN network are synchronized.

The OTN network performs a mapping and demapping at the edge of the network and constitutes an EEC function in the synchronization chain.

14.2. Control channel ESMC

Ethernet Synchronization Messaging Channel (ESMC) is composed of a Synchronization Status Message (SSM) encapsulated by the Organization-Specific Slow Protocol (OSSP) and Ethernet header (Figure 14.3).

```
 0 1 2 3 4 5 6 7 8 9 0 1 2 3 4 5 6 7 8 9 0 1 2 3 4 5 6 7 8 9 0 1
+-+-+-+-+-+-+-+-+-+-+-+-+-+-+-+-+-+-+-+-+-+-+-+-+-+-+-+-+-+-+-+-+
|                    Destination MAC Address                    |
+-+-+-+-+-+-+-+-+-+-+-+-+-+-+-+-+-+-+-+-+-+-+-+-+-+-+-+-+-+-+-+-+
|  Destination MAC Addr (cont)  |        Source MAC Addr        |
+-+-+-+-+-+-+-+-+-+-+-+-+-+-+-+-+-+-+-+-+-+-+-+-+-+-+-+-+-+-+-+-+
|                 Source MAC Address (continued)                |
+-+-+-+-+-+-+-+-+-+-+-+-+-+-+-+-+-+-+-+-+-+-+-+-+-+-+-+-+-+-+-+-+
|             Type              |    Subtype    |      OUI      |
+-+-+-+-+-+-+-+-+-+-+-+-+-+-+-+-+-+-+-+-+-+-+-+-+-+-+-+-+-+-+-+-+
|              OUI              |          OUI Subtype          |
+-+-+-+-+-+-+-+-+-+-+-+-+-+-+-+-+-+-+-+-+-+-+-+-+-+-+-+-+-+-+-+-+
| Vers. |C|                     Reserved                        |
+-+-+-+-+-+-+-+-+-+-+-+-+-+-+-+-+-+-+-+-+-+-+-+-+-+-+-+-+-+-+-+-+
|  Type: 0x01   |            Length             | Resvd |  QL   |
+-+-+-+-+-+-+-+-+-+-+-+-+-+-+-+-+-+-+-+-+-+-+-+-+-+-+-+-+-+-+-+-+
|                              FCS                              |
+-+-+-+-+-+-+-+-+-+-+-+-+-+-+-+-+-+-+-+-+-+-+-+-+-+-+-+-+-+-+-+-+
```

Figure 14.3. *Structure of the control channel ESMC*

14.2.1. *SSM message*

The SSM message is used to convey the quality of the source of the synchronization signal. SSM message involves in the selection of the interface that generates the reference clock.

The SMS message has a TLV (Type, Length and Value) structure, the value of the structure indicating the Quality Level (QL) of the clock:

– Type: this field encodes the message type on 1 byte (value 01 in hexadecimal);

– Length: this field encodes the size of the SSM messages on 2 bytes (value 00-04 in hexadecimal);

– Value: this field encodes the value of the quality level of the synchronization on 4 bits source associated with the Ethernet interface TE (Table 14.1):

- the value DNU (Do Not Use) indicates that the TE interface should not be used for the generation of reference clock T0,

- the value PRC indicates that the PRC clock is the source of the synchronization,

- the value SSU indicates that the SSU clock is the source of the synchronization,

- the value EEC1 is used when the SyncE mechanism is associated with the SDH network. EEC2 the value is used when the SyncE mechanism is associated with the SONET network.

QL PRC	QL SSU	QL EEC1	QL EEC2	DNU
0010	0100	1011	1010	1111

Table 14.1 *Quality Level (QL)*

New TLV structures can be introduced to identify the source of the synchronization signal to count the number of EEC and SSU clocks in the synchronization chain or to trace the EEC and SSU clocks constituting the synchronization chain.

The TLV structure "Primary Source Hop Count" is used to count the number of EEC and SSU clocks in the synchronization chain from the PRC in order to verify that the number of EEC clocks (SSU, respectively) does not exceed the value 60 (respectively, the value 10).

The TLV structure "Secondary Source Hop Count" is used to count the number of EEC clocks in the synchronization chain from the nearest SSU clock, to check that the number of clocks EEC does not exceed 20 between two SSU clocks.

The TLV structure "Primary Source Route" identifies the EEC and SSU clocks in the synchronization chain from the PRC in order to detect any loop. The TLV structure "Secondary Source Route" identifies the EEC clock in the synchronization chain from the nearest SSU clock to also detect any loop.

14.2.2. *OSSP header*

The SMS message is transmitted in a PDU (Packet Data Unit) OSSP structure whose header contains the following fields (Figure 14.3):

– Subtype: this field encodes on 1 byte the subtype of the OSSP protocol (value 0A in hexadecimal);

– OUI (Organizationally Unique Identifier): this field encodes the identifier assigned to the standard body ITU (International Telecommunication Union) (value 00-19-A7 in hexadecimal) on 3 bytes;

– OUI Subtype: this field encodes the subtype of ITU-OUI protocol (value 00-01 in hexadecimal) on 2 bytes;

– Version: this field encodes the version of ITU-OUI protocol (value 1 in hexadecimal) on 4 bits;

– C: this bit indicates if it is an information message (bit to ZERO) or an event message (bit to ONE):

- the event message is transmitted when there is a change in the transport synchronization,

- the information message is transmitted every second in the absence of synchronization transport change.

14.2.3. *Ethernet header*

The OSSP PDU structure is encapsulated with an Ethernet header (Figure 14.3):

– Destination MAC Address: this field encodes the multicast destination address (value 01-80-C2-00-00-02 in hexadecimal) on 6 bytes;

– Source MAC Address: this field encodes the unicast source address of the Ethernet frame on 6 bytes;

– Type: this field encodes the type of the encapsulated by the Ethernet header on 2 bytes (value 88-09 in hexadecimal for the OSSP PDU structure);

– FCS (Frame Check Sequence): this field encodes the value of the cyclic redundancy code on 4 bytes.

14.3. Network architecture based on EEC functions

SSM messages involve in the selection of the interface TE or T3 in order to generate the reference clock T0 independently, while avoiding the creation of synchronization loops.

In case of failure, the SSM message allows us to reconfigure the synchronization chain, EEC clocks seeking to recover the reference source with the highest hierarchical level.

14.3.1. *Ring network with two PRC sources*

The ring network, described in Figure 14.4, has two PRC sources.

Western PRC source synchronizes EEC#1 clock from T3 interface with priority 1. EEC#1 clock distributes the reference clock T0 on TE interfaces to EEC#2 and EEC#6 functions, indicating the quality level QL-PRC in an SSM message.

EEC#2 and EEC#3 clocks are synchronized with the signal received on TE interface with the priority 1, and generate the reference clock T0 for transmission on TE interfaces. The quality level QL-DNU is returned to avoid loops.

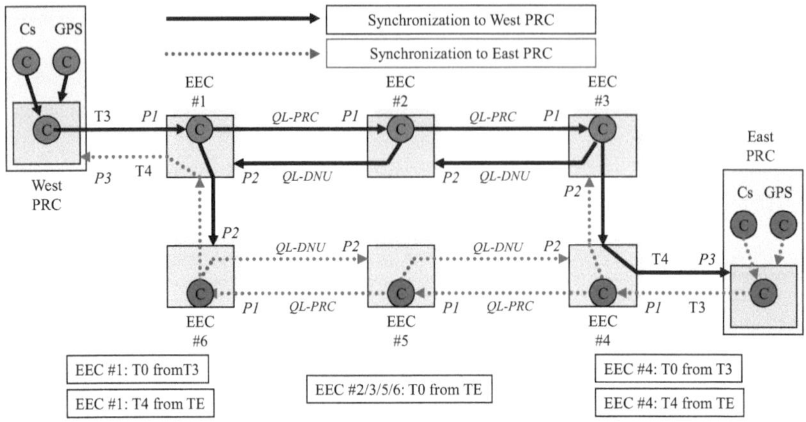

Figure 14.4. *Ring network with two PRC sources*

The same operations are carried out from the eastern PRC source that synchronizes EEC#4, EEC#5 and EEC#6 clocks.

EEC#4 clock retrieves the clock of the western PRC source and distributes it to the eastern PRC source on T4 interface.

Similarly, EEC#1 clock retrieves the clock of the eastern PRC source and distributes it to the western PRC source on T4 interface.

In case of failure of the sources from the Cesium and the GPS signal, the western PRC source (eastern, respectively) synchronizes the eastern PRC source (western, respectively).

In case of failure of an EEC function, e.g. the EEC#2 function, the synchronization chain reconfigures as follows (Figure 14.5):

– EEC #1 clock remains synchronized with western PRC source;

– EEC #4 and EEC#6 clocks remain synchronized with eastern PRC source;

– EEC #3 clock is synchronized with eastern PRC source;

– western PRC source can synchronize to the eastern PRC source;

– eastern PRC source cannot synchronize to the western PRC source.

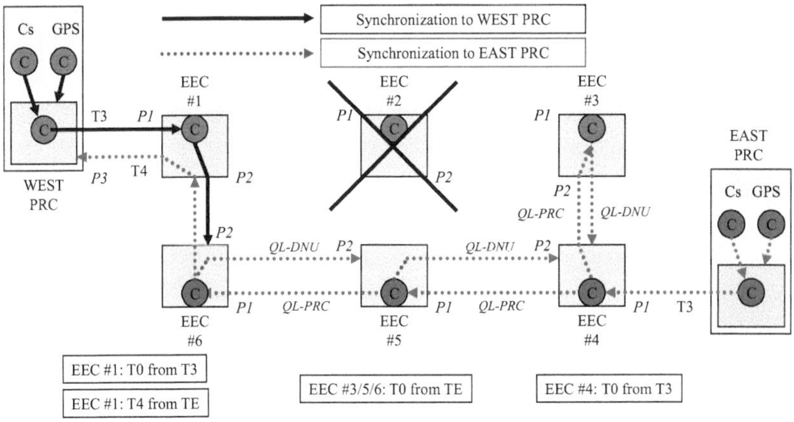

Figure 14.5. *Reconfiguration of the ring network with two PRC sources*

14.3.2. *Ring network with one SSU source and one SSU sink*

The ring network, described in Figure 14.6, is provided with an SSU source and an SSU sink.

EEC#1 to EEC#6 clocks are synchronized on SSU source. EEC#4 clock distributes the reference signal to the sink SSU.

EEC#1 clock is synchronized on SSU source from T3 interface with priority 1 and distributes the reference clock T0 on TE interfaces to EEC#2 and EEC#6 functions, indicating the quality level QL-PRC.

EEC#2 to EEC#6 clocks are synchronized on the received signal on TE interface with priority 1, and generates the reference clock T0 for the transmission on TE interfaces. The quality level QL-DNU is returned to avoid loops.

EEC#4 clock distributes the reference signal to the SSU sink on T4 interface, generated from the reference clock T0.

In case of failure of an EEC function, e.g. EEC#2 function, the initial synchronization chain (EEC#1/2/3/4/5/6) is reconfigured to reconstitute a new synchronization chain (EEC # 1/6/5/4/3) (Figure 14.7).

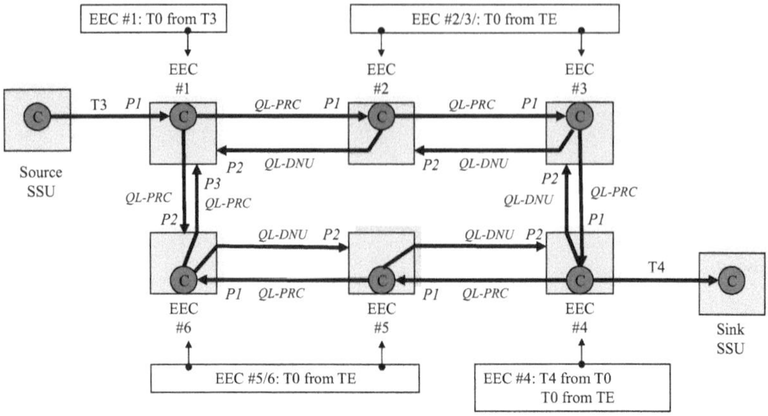

Figure 14.6. *Ring network with one SSU source and one SSU sink*

EEC#1 clock is synchronized with the SSU source from T3 interface with priority 1 and distributes the reference clock T0 on TE interface to EEC#6 function, indicating the quality level QL-PRC.

EEC#6/5/4/3 clocks are synchronized with the signal received on TE interface with the priority 2 and generate the reference clock T0 for transmission on the TE interfaces.

As before, the EEC#4 clock distributes the reference signal to the sink SSU on T4 interface generated from the reference clock T0.

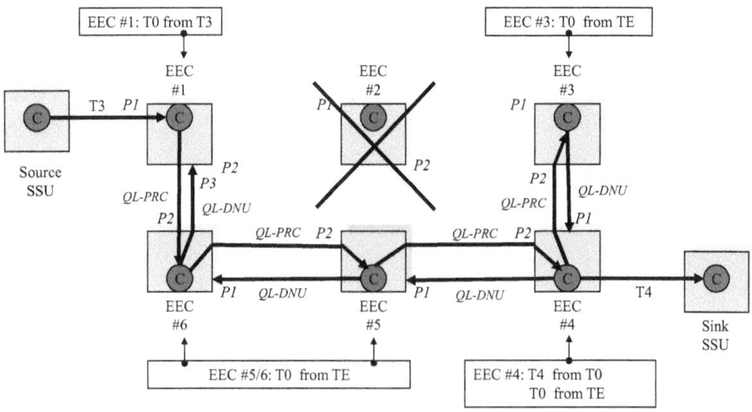

Figure 14.7. *Reconfiguration of the ring network with one SSU source and one SSU sink*

14.3.3. *Ring network with one SSU source and one SSU filter*

The ring network, described in Figure 14.8, is provided with an SSU source and an SSU filter.

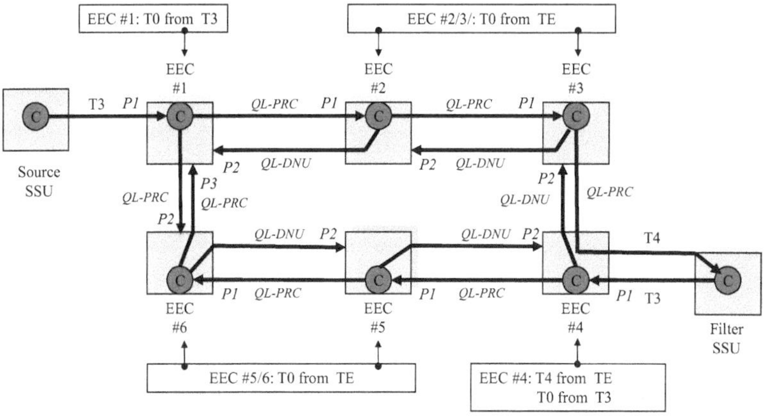

Figure 14.8. *Ring network with one SSU source and one SSU filter*

The SSU filter is inserted into the synchronization chain between the EEC#3 and EEC#4 functions.

EEC#4 clock distributes the reference signal to the SSU filter on T4 interface, generated from the received signal TE from the EEC#3 function.

EEC#4 clock synchronizes on the signal received from T3 interface with priority 1 and generates the reference clock T0 for transmission on TE interfaces.

In case of failure of an EEC function, e.g. EEC#2 function, as previously, the initial synchronization chain (EEC#1/2/3/4/5/6) is reconfigured to reconstitute a new synchronization chain (EEC#1/6/5/4/3) (Figure 14.9).

EEC#4 clock distributes the reference signal to the SSU filter on T4 interface, generated from the received signal TE from the EEC#5 function.

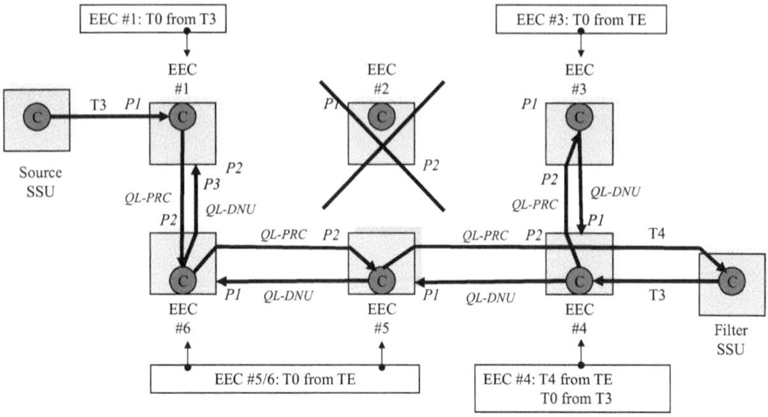

Figure 14.9. *Reconfiguration of the ring network with one SSU source and one SSU filter*

15

Synchronization IEEE 1588 Mechanism

15.1. Synchronization network architecture

IEEE 1588 mechanism defines Precision Time Protocol (PTP) and the characteristics of the entities that support this protocol in order to perform time synchronization.

The clocks are organized in a master/slave hierarchy, based on the information in the PTP messages.

Version 1 of the IEEE 1588 mechanism defines two types of entity. Ordinary Clock (OC) and Boundary Clock (BC).

The OC entity may be master if it is the time source or slave if it is synchronized with the master clock.

The BC entity has a slave port and multiple master ports:

– the slave port is synchronized to a master entity (OC) or to a master port of a BC entity;

– the master ports synchronize slave entities (OC) or the slave port of a BC entity.

BC entities can be located in a transmission loop. In this case, the lower-layer protocols (e.g. Ethernet or IP) must open the loop to establish a multicast tree.

The Grand Master (GM) clock located at the root of the time distribution tree is associated with master entity (OC) or a BC entity.

PTP protocol is exchanged between a master port (master entity (OC) or master port of a BC entity) and slave ports (slave ports of a BC entity or slave entities (OC)) (Figure 15.1).

PTP messages allow for the measurement of the transfer time between a slave port and the associated master port in order to synchronize the slave port.

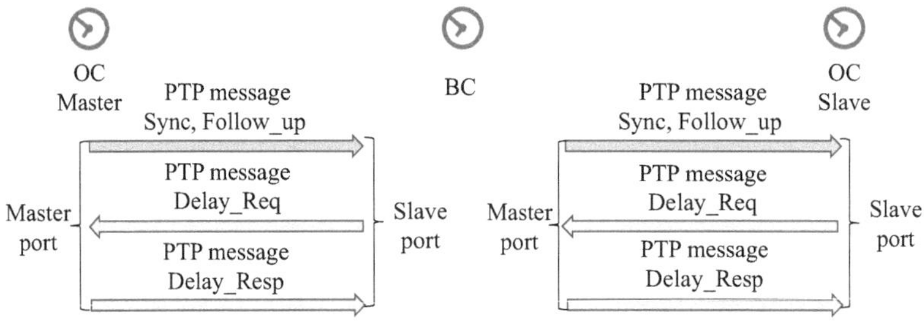

Figure 15.1. *Time synchronization architecture*

Version 2 of IEEE 1588 mechanism introduces the Transparent Clock (TC):

– entity E2E (End-to-End) TC measures the transfer time in the TC entity for the PTP messages Sync and Delay_Req;

– entity P2P (Peer-to-Peer) TC measures the transfer time in the TC entity for the PTP messages and the propagation time to the adjacent entities.

TC entities do not complete PTP protocol. They modify the content of the messages Follow_up or Delay_Resp by placing the measurements.

Time synchronization of version 1 is based on the symmetry of transfer time of the PTP messages between, firstly, the master port and the slave ports for the downlink direction, and, secondly, between the slave port and the master ports for the uplink direction.

Entity E2E TC will correct the asymmetry of transfer times of PTP messages in the various entities E2E TC (Figure 15.2):

– the measurement value of the transfer time of the PTP message Sync is inserted into the PTP messages Follow_up;

– the measurement value of the transfer time of the PTP message Delay_Req is inserted into the PTP message Delay_Resp.

Entities E2E TC are not necessarily synchronized to the master entity (OC). However, a frequency difference between the master entity OC and the entity E2E TC can introduce an error in the calculation of the transfer time.

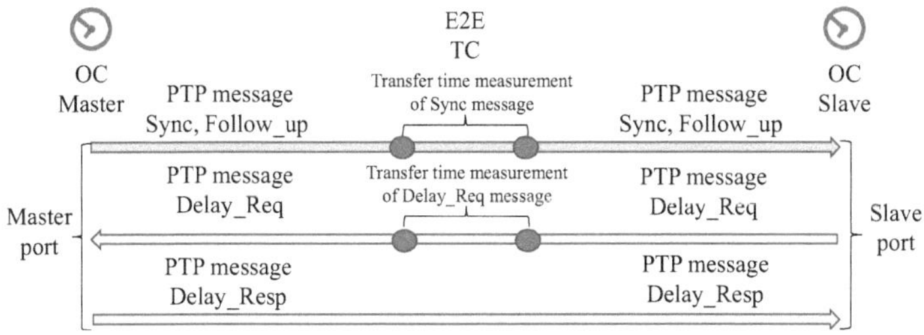

Figure 15.2. *Transfer time measurement*

Time synchronization of version 1 is based on measuring the propagation time between the master port and the slave ports. The introduction of the entity P2P TC allows a calculation of the link propagation time (Figure 15.3).

Calculating the propagation time between two adjacent entities is done when it is not carried end-to-end between the master and slave entities (OC).

The measurement value of the link propagation time is inserted into the PTP message Follow_up.

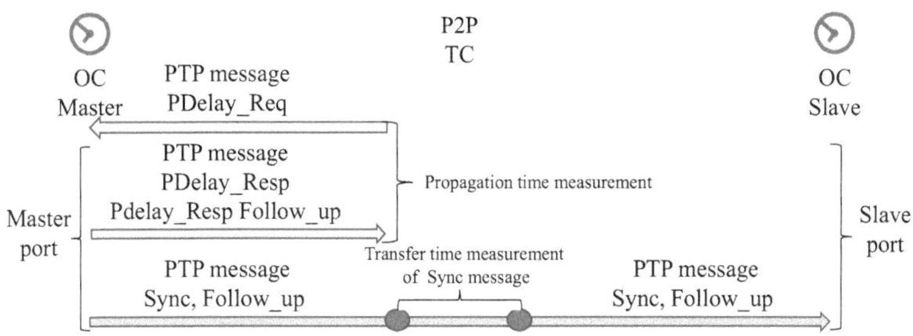

Figure 15.3. *Propagation time measurement*

15.2. Time synchronization principles

15.2.1. *Version 1 of the time synchronization*

The principle of the time synchronization is described in Figure 15.4 for the version 1 of IEEE 1588 mechanism.

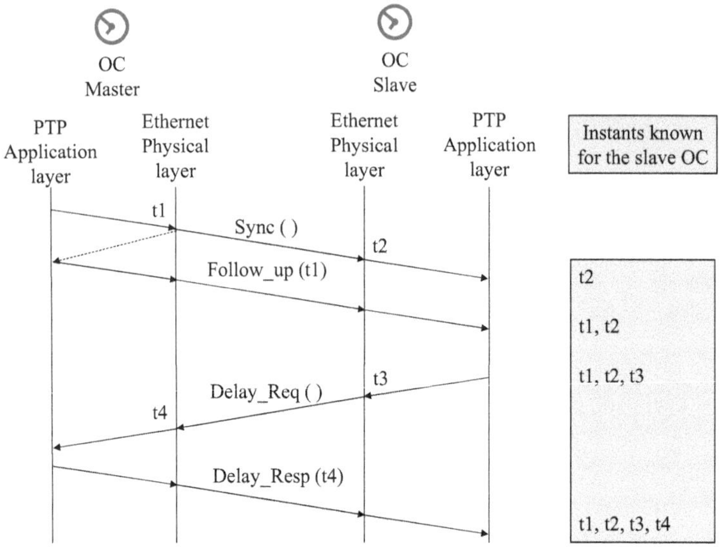

Figure 15.4. *Version 1 of the time synchronization*

PTP message Sync transmitted by the master entity (OC) allows the slave entity (OC) to store the date t2 at which the PTP message Sync is received.

In the case where the PTP message follow-up is not used (one-step mechanism), the PTP message Sync contains the time t1 at which the PTP message Sync was transmitted. The value t1 corresponds to the date the PTP Sync message was created by the application layer.

The value t1 is refined in the PTP message Follow_up (two-step mechanism). The new value of t1 is the measurement made by the master entity (OC) at the physical layer of the Ethernet interface.

The value t2 corresponds to the measurement made by the slave entity (OC) at the physical layer of the Ethernet interface.

The slave entity (OC) can then calculate the value MS_Difference:

MS_Difference = t2 – t1 = Offset + MS_Delay [15.1]

Offset: time difference between the master and the slave clocks

MS_Delay: propagation time between the master and the slave entities.

The slave entity (OC) sends the PTP message Delay_Req at the date t3 that corresponds to the measurement made by the slave entity (OC) at the physical layer of the Ethernet interface.

The master entity (OC) determines the date t4 on which the PTP message Delay_Req was received. The date t4 corresponds to the measurement performed by the master entity (OC) at the physical layer of the Ethernet interface.

The master entity (OC) transmits the PTP messages Delay_Resp with the value t4 to the slave entity (OC), which calculates the value SM_Difference.

SM_Difference = t4 – t3 = SM_Delay – Offset [15.2]

SM_Delay: propagation time between the slave and the master entities

From equations [15.1] and [15.2], the slave entity (OC) gets the value of the time offset between the master and slave entities and the average propagation time one_way_delay between these two entities, assuming that the delay for each transmission direction is identical.

Offset = [(MS_Difference – SM_Difference) – (MS_Delay – SM_Delay)]/2

Offset = (MS_Difference – SM_Difference)/2

one_way_delay = (MS_Delay + SM_Delay)/2

one_way_delay = (MS_Difference + SM_Difference)/2

From the calculated offset value, the slave entity (OC) can update its clock. Note that asymmetric delay for each direction of transmission results in an error of setting the time of the slave clock.

15.2.2. *Version 2 of the time synchronization*

The principle of time synchronization is described in Figure 15.5 for the version 2 of IEEE 1588 mechanism, the correction relating only to the transfer time of different entities.

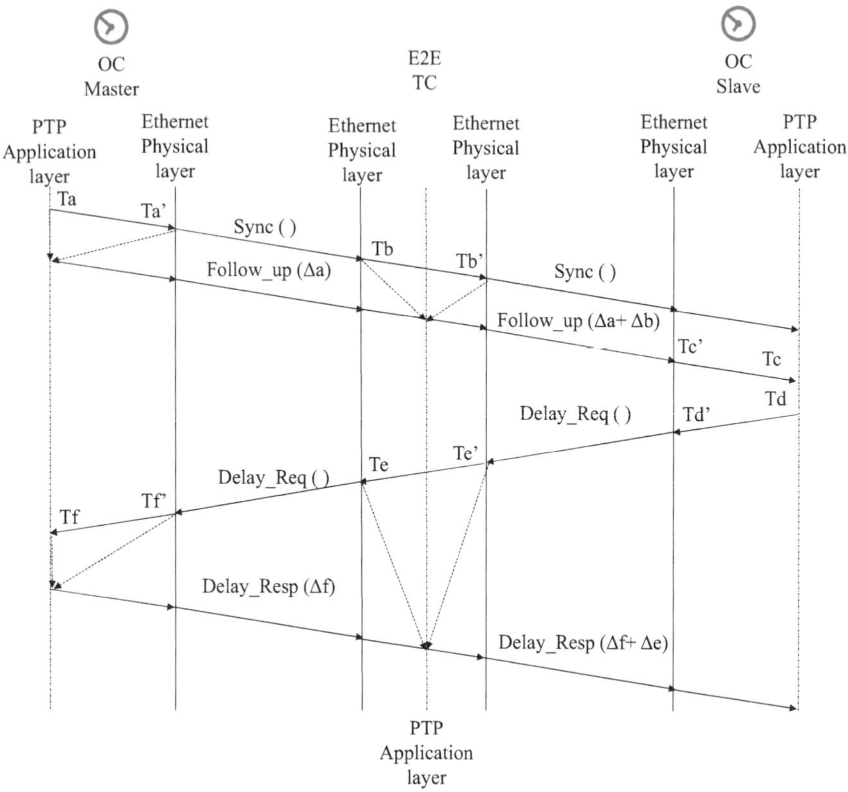

Figure 15.5. *Version 2 of the time synchronization*

The master entity (OC) transmits in the correction field of the PTP message Follow_up the value of transfer time (Δa) for the PTP message Sync, corresponding to the difference of time measurements at the application layer PTP (Ta) and the physical layer (Ta').

Upon receiving the PTP message Sync, the entity E2E TC triggers the measurement of the transfer time (Δb) at the physical layer based on the reception and the transmission of the PTP message Sync.

The value of the transfer time (Δb) is added to that (Δa) transmitted by the master entity (OC) in the correction field of the PTP message Follow_up.

Upon receiving the PTP message Sync, the slave entity (OC) calculates the value of the transfer time (Δc), corresponding to the difference of time measurements at the application layer (Tc) and the physical layer (Tc').

Upon receiving the PTP message Follow_Up, the slave entity (OC) retrieves the value of the transfer time (Δa + Δb) of the master entities OC and E2E TC.

The slave entity (OC) measures the transfer time (Δd) of the PTP message Delay_Req, corresponding to the difference of time measurements at the application layer (Td) and the physical layer (Td').

Upon receiving the PTP message Delay_Req, the entity E2E TC triggers the measurement of the transfer time at the physical layer, based on the reception and the transmission of the PTP message Delay_Req.

Upon receiving the PTP message Delay_Req, the master entity (OC) calculates the value of the transfer time (Δf), corresponding to the difference of time measurements at the application layer (Tf) and the physical layer (Tf').

The master entity (OC) transmits in the correction field of the PTP message Delay_Resp the value of transfer time (Δf) of the PTP message Delay_Req.

The entity E2E TC adds the value of the transfer time (Δe) of the PTP message Delay_Req to that (Δf) transmitted by the master entity (OC) in the correction field of the PTP message Delay_Resp.

After the exchange of PTP messages, the slave entity (OC) can calculate the time difference related to the asymmetry of the transfer times for the various entities between each transmission direction.

Calculating the propagation time between two adjacent entities (DelayRequestor and DelayResponder) is described in Figure 15.6 for version 2 of IEEE 1588 mechanism.

The entity DelayRequestor sends to the adjacent entity DelayResponder the PTP message PDelay_Req and measures, at the physical layer, the time T1 of the transmission of the message.

The entity DelayResponder determines the date T2 of the reception of the PTP message PDelay_Req measured at the physical layer that transmits to the entity DelayRequestor in the PTP message PDelay_Resp.

The entity DelayResponder transmits the date T3 of the transmission of the PTP message PDelay_Resp in the PTP message PDelay_Resp_FollowUp.

The entity DelayRequestor determines the date T4 of the reception of the PTP message PDelay_Resp, measured at the physical layer.

The average time of the propagation between the entities DelayRequestor and DelayResponder is [(T2 - T1) + (T4 - T3)] / 2.

The value of the propagation time between two adjacent entities is added to the correction field of the PTP message Follow_up.

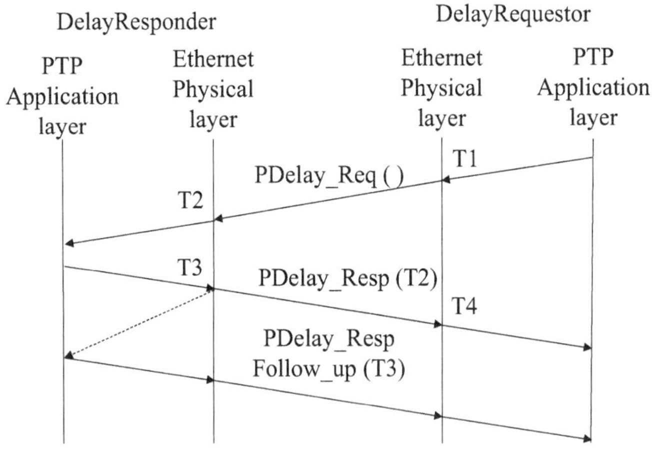

Figure 15.6. *Propagation time calculation between two entities*

15.3. PTP messages

15.3.1. *Message format*

The PTP message includes a header common to all messages and a specific body to each message.

The header of the PTP message has a fixed length of 34 bytes and consists of the following different fields (Table 15.1):

– transportSpecific: this field, encoded on 4 bits, can be used by the transport protocol of the PTP message;

– messageType: this field encodes on 4 bits the type of the PTP message (Sync=0x0, Delay_Req=0x1, PDelay_Req=0x2, PDelay_Resp=0x3, Follow_up=0x8, Delay_Resp=0x9, PDelay_Resp_Follow_up=0xA, Announce=0xB, Signaling=0xC, Management=0xD);

– versionPTP: this field encodes on 4 bits the version of the PTP protocol (value of 1 or 2);

– domainNumber: this field, encoded on 1 byte, identifies the domain to which the PTP message belongs. A domain is a logical set of clocks that synchronize each other. Clocks that belong to different domains are not necessarily synchronized;

– messageLength: this field encodes on 2 byte the size of the PTP message;

Bits								Bytes
7	6	5	4	3	2	1	0	
TransportSpecific				messageType				1
Reserved				versionPTP				1
MessageLength								2
DomainNumber								1
Reserved								1
FlagsField								2
CorrectionField								8
Reserved								4
SourcePortIdentity								10
SequenceId								2
ControlField								1
LogMeanMessageInterval								1

Table 15.1. *Format of the common header*

– flagsField: this field encodes on 2 bytes the characteristics for using PTP protocol (Alternate Master, two step, Unicast, Profile Specific 1, Profile Specific 2, Security, Leap Indicator 61, Leap Indicator 59, UTC Reasonable, Timescale, Time Traceable, Frequency Traceable);

– correctionField: this field encodes on 8 bytes the value, in nanoseconds, of the transfer time (TC entities) and the propagation time (entity P2P TC);

– sourcePortIdentity: this field, encoded on 10 bytes, identifies the origin of the port that issued the message,

– sequenceId: this field encodes on 2 bytes the sequence number of the message;

– controlField: this field, encoded on 1 byte, has a value that depends on the type of message. It is used only for PTP messages of version 1;

– logMeanMessageInterval: this field encodes on 1 byte the value, in seconds, of the time interval of transmission of the message. For example, for a time interval of 0.125 s, the value of this field is equal to –3 ($0,125 = 2^{-3}$).

The body of the messages Sync and Delay_Req contains a single field, originTimestamp, coded on 10 bytes (Table 15.2).

This field contains the value of the date t1 of the transmission of the message Sync (one-step mechanism) and the message Delay_Req, measured at the application layer.

In the case of using the message Follow_Up (two-step mechanism), the value of this field is zero for the message Sync.

Bits								Bytes
7	6	5	4	3	2	1	0	
Header								34
OriginTimestamp								10

Table 15.2. *Format of the messages Sync and Delay_Req*

The body of the message Follow_Up contains a single field: preciseOriginTimestamp (Table 15.3).

This field encodes on 10 bits the value of the date t1 of the transmission of the message Sync measured at the physical layer of the master port.

Bits								Bytes
7	6	5	4	3	2	1	0	
Header								34
PreciseOriginTimestamp								10

Table 15.3. *Format of the message Follow_up*

The body of the message Delay_Resp contains two fields: receiveTimestamp and requestingPortIdentity (Table 15.4):

– receiveTimestamp: this field encodes on 10 bytes the arrival date t4 of the message Delay_Req measured at the master port;

– requestingPortIdentity: this field encodes on 10 bytes the identity of the slave port that transmitted the message Delay_Req.

Bits								Bytes
7	6	5	4	3	2	1	0	
Header								34
ReceiveTimestamp								10
RequestingPortIdentity								10

Table 15.4. *Format of the message Delay_Resp*

The body of the message PDelay_Req contains two fields: originTimestamp and reserved (Table 15.5):

– originTimestamp: this field encodes on 10 bits the value of the time of the entity DelayRequestor;

– reserved: this field, encoded on 10 bytes, is introduced so that the messages PDelay_Req and PDelay_Resp have the same size in order to avoid an error related to the asymmetry of sizes.

Bits								Bytes
7	6	5	4	3	2	1	0	
Header								34
OriginTimestamp								10
Reserved								10

Table 15.5. *Format of the message PDelay_Req*

The body of the message PDelay_Resp contains two fields: receiveReceiptTimestamp and requestingPortIdentity (Table 15.6):

– receiveReceiptTimestamp: this field encodes on 10 bits the value of the arrival time T2 of the message PDelay_Req measured at the physical layer of the master port;

– requestingPortIdentity: this field encodes on 10 bits the identity of entity DelayRequestor that transmitted the message PDelay_Req.

Bits								Bytes
7	6	5	4	3	2	1	0	
Header								34
ReceiveReceiptTimestamp								10
RequestingPortIdentity								10

Table 15.6. *Format of the message PDelay_Resp*

The body of the message PDelay_Resp_Follow_up contains two fields: responseOriginTimestamp and requestingPortIdentity (Table 15.7):

– responseOriginTimestamp: this field encodes on 10 bits the value of the time T3 of the transmission of the message PDelay_Resp measured at the physical layer of the entity DelayResponder.

Bits								Bytes
7	6	5	4	3	2	1	0	
Header								34
ResponseOriginTimestamp								10
RequestingPortIdentity								10

Table 15.7. *Format of the message PDelay_Resp_Follow_up*

The message Announce was introduced in version 2 of the PTP protocol to determine the best source GM via the Best Master Clock Algorithm (BMCA) and the status (master or slave) of the port.

The message Sync was used in version 1 of the PTP protocol for this function.

The body constitution of the message Announce is provided in Table 15.8.

– currentUTCOffset: this field encodes on 2 bytes the offset from the Universal Time Coordinated (UTC);

– timeSource: this field encodes on 1 byte the type of GM clock used by the time reference (ATOMIC_CLOCK, GPS, TERRESTRIAL_RADIO and so on);

– stepsRemoved: this field encodes on 2 bytes the number of hops to reach the master clock GM. This field is used by the algorithm BMCA;

– grandmasterPortIdentity: this field encodes on 10 bytes the identity of the port of the master clock GM. This field is used by the algorithm BMCA;

– grandmasterClockQuality: this field encodes of 4 bytes the quality of master clock GM. This field is used by the algorithm BMCA. This field consists of three parameters:

- the parameter clock Class encodes the nature of the time reference;
- the parameter clockAccuracy encodes the level of the clock accuracy;
- the parameter offsetScaledLogVariance encodes the level of the clock stability;

– grandmasterPriority1: this field encodes on 1 byte the first level of priority. The value is usually set to 128. This field is used by the algorithm BMCA;

– grandmasterPriority2: this field encodes on 1 byte the second level of priority. This field is used by the algorithm BMCA.

Bits								Bytes
7	6	5	4	3	2	1	0	
En-tête								34
OriginTimestamp								10
CurrentUTCOffset								2
Reserved								1
TimeSource								1
StepsRemoved								2
GrandmasterPortIdentity								10
GrandmasterClockQuality								4
grandmasterPriority1								1
grandmasterPriority2								1
ParentPortIdentity								10

Table 15.8. *Format of the message Announce*

15.4.2. *Message transport*

PTP messages are transmitted in IPv4 or IPv6 multicast packets:

– IPv4 destination address: 224.0.1.129 for all messages except the messages PDelay_Req, PDelay_Resp, Pdelay_Resp_Follow_up;

– IPv4 destination address: 224.0.0.107 for the messages PDelay_Req, PDelay_Resp, Pdelay_Resp_Follow_up;

– IPv6 destination address: FF0x::181 for all messages except the messages PDelay_Req, PDelay_Resp, Pdelay_Resp_Follow_up;

– IPv6 destination address: FF02::6B for the messages PDelay_Req, PDelay_Resp, Pdelay_Resp_Follow_up.

PTP messages are encapsulated by a UDP (User Datagram Protocol) header:

– destination port number: 319 for the messages Sync, Delay_Req, Pdelay_Req, Pdelay_Resp;

– destination port number: 320 for other PTP messages.

PTP messages can also be transmitted directly in Ethernet multicast frames:

– MAC destination address: 01-1B-19-00-00-00 for all messages except the messages PDelay_Req, PDelay_Resp, Pdelay_Resp_Follow_up;

– MAC destination address: 01-80-C2-00-00-0E for the messages PDelay_Req, PDelay_Resp, Pdelay_Resp_Follow_up;

– value of the field type: 88F7 (in hexadecimal).

Bibliography

Chapter 1: Mobile Network – Architecture

[3GPP TS 23.401] General Packet Radio Service (GPRS) enhancements for Evolved Universal Terrestrial Radio Access Network (E-UTRAN) access

[NETMANIAS] LTE Network Architecture: Basic

Chapter 2: Mobile Network – Signaling Protocols

[3GPP TS 24.301] Non-Access-Stratum (NAS) protocol for Evolved Packet System (EPS)

[3GPP TS 36.331] Radio Resource Control (RRC): Protocol specification

[3GPP TS 36.413] S1 Application Protocol (S1AP)

[3GPP TS 36.423] X2 Application protocol (X2AP)

[3GPP TS 29.274] Evolved General Packet Radio Service (GPRS) Tunneling Protocol for Control plane (GTPv2-C)

[IETF RFC 3588] Diameter Base Protocol

[3GPP TS 29.272] Mobility Management Entity (MME) and Serving GPRS Support Node (SGSN) related interfaces based on Diameter protocol

[3GPP TS 29.212] Policy and Charging Control (PCC); Reference points

[3GPP TS 29.215] Policy and Charging Control (PCC) over S9 reference point; Stage 3

Chapter 3: Mobile Network – Procedures

[3GPP TS 23.401] General Packet Radio Service (GPRS) enhancements for Evolved Universal Terrestrial Radio Access Network (E-UTRAN) access

[NETMANIAS] EMM Procedure 1. Initial Attach

[NETMANIAS] EMM Procedure 2. Initial Attach

[NETMANIAS] EMM Procedure 4. Service Request

Chapter 4: Transport Network – MPLS-VPN Technology

[IETF-RFC 3031] Multiprotocol Label Switching Architecture

[IETF-RFC 3032] MPLS Label Stack Encoding

[IETF-RFC 5036] LDP Specification

[IETF-RFC 4364] BGP/MPLS IP Virtual Private Networks (VPNs)

[JUNIPER] RFC 2547bis: BGP/MPLS VPN Fundamentals

Chapter 5: Transport Network – VPLS Technology

[IETF-RFC 4448] Encapsulation Methods for Transport of Ethernet over MPLS Networks

[IETF-RFC 4447] Pseudowire Setup and Maintenance Using the Label Distribution Protocol (LDP)

[IETF-RFC 4761] Virtual Private LAN Service (VPLS) Using BGP for Auto-Discovery and Signaling

[IETF-RFC 4762] Virtual Private LAN Service (VPLS) Using Label Distribution Protocol (LDP) Signaling

[HUAWEI] VPLS Technology White Paper

Chapter 6: Transport Network – OTN Technology

[ITU-T-G.872] Architecture of optical transport networks

[ITU-T-G.694.1] Spectral grids for WDM applications: DWDM frequency grid

[IUT-T-G.694.2] Spectral grids for WDM applications: CWDM wavelength grid

[ITU-T-G.709] Interfaces for the optical transport network

[ITU-T-G.7041] Generic framing procedure

[EXFO] OTN (G.709) Reference Guide

[PMC-SIERRA] A Tutorial on ITU-T G.709 Optical Transport Networks (OTN) – Steve Gorshe

[PMC-SIERRA] Generic Framing Procedure (GFP) – Steve Gorshe

Chapter 7: Quality of Service – Principles

[ITU-T-Y.1541] Network performance objectives for IP-based services

[ITU-T-G.107] The E-model: a computational model for use in transmission planning

Chapter 8: Quality of Service – Mobile Network

[3GPP TS23.203] Policy and charging control architecture

[NETMANIAS] LTE QoS: SDF and EPS Bearer QoS

Chapter 9: Quality of Service – Transport Network

[IETF-RFC 2474] Definition of the Differentiated Services Field (DS Field) in the IPv4 and IPv6 Headers

[IETF-RFC 2475] An Architecture for Differentiated Services

[IETF-RFC 2597] Assured Forwarding PHB Group

[IETF-RFC 3246] An Expedited Forwarding PHB (Per-Hop Behavior)

[IETF-RFC 4594] Configuration Guidelines for DiffServ Service Classes

[IETF-RFC 2697] A Single Rate Three Color Marker

[IETF-RFC 2698] A Two Rate Three Color Marker

[IETF-RFC 3270] Multi-Protocol Label Switching (MPLS) Support of Differentiated Services

[IEEE-802.1P] Annex G – Priority

Chapter 10: Security – Principles

[NIST FIPS PUB 197] Advanced Encryption Standard (AES)

[NIST FIPS PUB 198-1] The Keyed-Hash Message Authentication Code (HMAC)

[NIST FIPS PUB 180-4] Secure Hash Standard (SHS)

[RFC 2104] HMAC: Keyed-Hashing for Message Authentication

[RFC 1321] The MD5 Message-Digest Algorithm

[RFC 2631] Diffie-Hellman Key Agreement Method

[RFC 3526] More Modular Exponential (MODP) Diffie-Hellman groups for Internet Key Exchange (IKE)

Chapter 11: Security – Mobile Network

[3GPP TS 33.401] 3GPP System Architecture Evolution (SAE); Security architecture

[3GPP TS 24.301] Non-Access-Stratum (NAS) protocol for Evolved Packet System (EPS)

[NETMANIAS] LTE Security I: LTE Security Concept and LTE Authentication

[NETMANIAS] LTE Security II: NAS and AS Security

Chapter 12: Security – Transport Network

[RFC 4301] Security Architecture for the Internet Protocol

[RFC 4302] IP Authentication Header

[RFC 4303] IP Encapsulating Security Payload (ESP)

[NETMANIAS] LTE Security I: LTE Security Concept and LTE Authentication

[RFC 5996] Internet Key Exchange Protocol Version 2 (IKEv2)

[3GPP 3310] Network Domain Security (NDS); Authentication Framework (AF)

Chapter 13: Synchronization – Principles

[ITU-T G.810] Definitions and terminology for synchronization networks

[ITU-T G.8260] Definitions and terminology for synchronization in packet networks

[ITU-T G.8261] Timing and synchronization aspects in packet networks

[ITU-T G.8261-1] Packet delay variation network limits applicable to packet-based methods (Frequency synchronization)

[ITU-T G.8262] Timing characteristics of a synchronous Ethernet equipment slave clock

[3GPP TS 24.011] Point-to-Point (PP) Short Message Service (SMS) support on mobile radio interface

[3GPP TS 36.401] Evolved Universal Terrestrial Radio Access Network (E-UTRAN); Architecture description

[3GPP TS 36.104] Evolved Universal Terrestrial Radio Access (E-UTRA); Base Station (BS) radio transmission and reception

[3GPP TS 36.133] Evolved Universal Terrestrial Radio Access (E-UTRA); Requirements for support of radio resource management

[3GPP TS 36.814] Evolved Universal Terrestrial Radio Access (E-UTRA); Further advancements for E-UTRA physical layer aspects"

Chapter 14: Synchronization – SyncE Mechanism

[ITU-T G.8264] Distribution of timing information through packet networks

Chapter 15: Synchronization – IEEE 1588 Mechanism

[IEEE 1588] Precision Clock Synchronization Protocol for Networked Measurement and Control Systems

[AGILENT] IEEE 1588: An Update On The Standard And Its Application – John C. Eidson

[CALNEX SOLUTIONS] Implementing IEEE 1588v2 for use in the mobile backhaul

Index

B

C

D

E

H

I

N

O

P

Q

R

S

T

U

V

W, X

FSC
www.fsc.org
MIX
Papier | Fördert
gute Waldnutzung
FSC® C083411

Kolibri 360 GmbH
Ferdinand-Jühlke-Straße 7
99095 Erfurt, Deutschland
produktsicherheit@kolibri360.de